Residential Electrical Wiring

Residential Electrical Wiring

REX MILLER

Professor of Industrial Arts and Vocational Education
State University College
Buffalo, New York

Bennett & McKnight
a division of
Glencoe Publishing Company
Encino, California

Published by Bennett & McKnight

Copyright transferred to Glencoe Publishing Company, a division of Macmillan, Inc.

Send all inquiries to:
Glencoe Publishing Company
17337 Ventura Boulevard
Encino, California 91316

Library of Congress Catalog Card Number: 80-67830

Printed in the United States of America

ISBN 0-02-665620-5 (Student Text)
ISBN 0-02-665640-X (Student Guide)
ISBN 0-02-665630-2 (Teacher's Guide)

(Previously ISBN 87002-331-4)

3 4 5   89 88 87 86

Cover design and drawings: Jim Kasprzyk

Preface

The purpose of *Residential Electrical Wiring* is to aid you in the study of house circuits and basic electricity. It presents information that will enable you to understand and work with electricity in the residential sector of the field.

The book was designed for students in vocational classes as well as for electricians enrolled in apprenticeship courses who desire training in residential wiring Although it contains some theory of the principles of electricity, the main emphasis of the text is on the practical, everyday applications of those principles.

Many illustrations are included to show the variety of parts and techniques presently used. Obviously, not all related problems can be presented here since there is a great deal of ingenuity required of the worker on the job. For standard procedure, however, the National Electrical Code *Handbook* does give a guide to the number of wires and the types of equipment safe to install in a given type of location. The *Handbook* should be a part of your tool kit.

As you know, it is not possible to learn to swim without first trying yourself in the water. Neither is it possible to learn all there is to know about the electrical field by reading about it. You must be willing to get some practical experience and devote some time to the development of skills related to the job.

It is the desire of the author that this book will provide the basic information needed by those really interested in going into the field of electricity.

Dedication

To Frank Navara, Jr.

Acknowledgments

As is the case with any book, many people have given generously of their time in assisting the author. Their efforts and suggestions have been of great value. I would like to take this opportunity to express my appreciation.

Many manufacturers of electrical parts and equipment have contributed illustrations, written materials, and suggestions for inclusion in this book. They have done so with a great deal of professionalism.

Manufacturers have been identified with each of the illustrations furnished. The following list gives their complete names. These manufacturers and agencies are but a sampling of the many who make the generation, installation, and use of electricity as efficient and safe as possible. In addition, the technical data furnished by these companies and agencies are much appreciated. Without their assistance, this project could have had neither appeal nor objectivity.

Appleton Electric Company, Chicago, Illinois

Bernzomatic Corporation, Rochester, New York

Brown & Sharpe Manufacturing Company, Providence, Rhode Island

Bryant Electric Company, Bridgeport, Connecticut

Buchanan Electric Products Division, Union, New Jersey

Canadian Standards Association, Rexdale, Ontario, Canada

Central Illinois Light Company (CILCO), Peoria, Illinois

Circle F Industries, Trenton, New Jersey

Edison Electric Institute, New York, New York

Eller Manufacturing Co., Brooklyn, New York

Emerson Electric Company, St. Louis, Missouri

General Electric Company, Providence, Rhode Island

Greenlee Tool Company, Rockford, Illinois

Harvey Hubbell, Inc., Bridgeport, Connecticut

Heath Company, Benton Harbor, Michigan

Honeywell, Minneapolis, Minnesota

Ideal Industries, Inc., Sycamore, Illinois

Misener Manufacturing Company, Inc., Syracuse, New York

National Fire Protection Association, Boston, Massachusetts

National Safety Council, Chicago, Illinois

New York Board of Fire Underwriters, New York, New York

New York State Power Authority, Niagara Falls, New York

Niagara Mohawk Power Corporation, Buffalo, New York

NuTone Division of Scovill, Cincinnati, Ohio

Ohmite, Skokie, Illinois

Onan, Minneapolis, Minnesota

Pass and Seymour, Inc., Syracuse, New York

Rural Electrification Authority (REA), Washington, D.C.

Sangamo Electric Company, Springfield, Illinois

Sears, Roebuck & Company, Chicago, Illinois

Seatek Company, Riverside, Connecticut

Slater Electric, Inc., Glen Cove, New York

Square D Company, Lexington, Kentucky

Superior Electric Company, Bristol, Connecticut

Thomas & Betts Company, Elizabeth, New Jersey

3M, Electro-Products Division, St. Paul, Minnesota

Underwriters' Laboratories, Inc., Chicago, Illinois

Union Insulating Company, Parkersburg, West Virginia

U. S. Department of Labor

Weston Electrical Instruments Company, Newark, New Jersey

Table of Contents

PREFACE 5

ACKNOWLEDGMENTS 6

CHAPTER 1: ELECTRICITY AND ELECTRICAL CIRCUITS 11
    What is Electricity?, 11
    Metric Measurement In Electricity, 11
    Elements and Atoms, 11
    Electrical Materials, 13
    Generating Electricity, 15
    Voltage and Current, 18
    Ohm's Law, 24
    Power, 25
    Measuring Electricity, 27
    Controlling Electricity, 32
    Resistors, 37
    Resistor Color Code, 38

CHAPTER 2: PRODUCING ELECTRICITY 43
    Power Plant Operations—AC, 45
    Falling-Water Generators, 52
    Nuclear Power Plants, 54
    Fossil-Fuel Power Generators, 56

CHAPTER 3: DISTRIBUTING ELECTRICITY 61
    Urban and Suburban Distribution, 61
    Rural Electricity, 68
    Installation of Service by Power Company, 70

CHAPTER 4: ELECTRICAL SAFETY 79
    Temporary Wiring, 79
    Temporary Lighting, 79
    Temporary Cords, 81
    Ladders, 81
    Fatal Current, 84
    General Safety Precautions, 85
    What To Do For Victims, 85
    Typical Shock Hazards with Extension Cords, Plugs, and
      Receptacles, 87
    Grounding Adapters, 95
    Safety Devices, 99

CHAPTER 5: WIRING TO THE ELECTRICAL CODE REQUIREMENTS 106
  National Electrical Code (NEC), 106
  Underwriters' Laboratories, 107
  Canadian Standards Association (CSA), 109
  Wiring According to the National Electrical Code, 110

CHAPTER 6: PLANNING 123
  Figuring Circuit Load, 123
  Interpreting Requirements, 125
  Typical Outlet Requirements for a Home, 127

CHAPTER 7: EQUIPMENT AND TOOLS FOR WIRING 145
  Basic Tools, 145
  Tools for General Use, 153

CHAPTER 8: HOUSE SERVICE AND CIRCUITS 160
  Service Entrance, 161
  Distribution Panels, 169
  Romex Cable, 171
  Wire Size, 172
  Planning the Right Size Service and Circuits, 172
  Kilowatt-Hour Meters, 193

CHAPTER 9: INSTALLING ROMEX 196
  Romex, 196
  BX, 197
  Fixtures and Splices, 209
  Number of Conductors in a Box, 216
  Aluminum Connections, 220
  Circuit Protective Devices, 222

CHAPTER 10: LOW-VOLTAGE AND LIGHTING CIRCUITS 226
  Low-Voltage Remote Control, 226
  Low-Voltage Wiring, 245
  Lighting Controls, 248

CHAPTER 11: WIRING WITH BX, EMT, FLEXIBLE, PLASTIC, AND RIGID
  CONDUIT 257
  BX Cable, 257
  Fittings, 258
  BX Cable and the Code, 262
  Installing Electric Service with Conduit, 264
  Conduits and the Code, 283
  Flexible Conduit, 287

CHAPTER 12: REMODELING WIRING 291
  Circuit Testing, 292
  Adding Another Outlet, 292
  Wall Switch Added to Control Outlet, 295

Adding an Outlet along the Wall, 296
Running a Wire around a Door, 298
Running a Wire through the Floor, 299
How to Fish a Wire, 301
Attic Installation, 301
Installing a Secure Box, 302
Underground Wiring, 304
Covers and Canopies, 306
Outdoor Lighting, 307

CHAPTER 13: APPLIANCES, OVENS, RANGES, COOKING TOPS, AND AIR
CONDITIONERS 310
Electric Ranges, 310
Dryer, 317
Microwave Ovens, 318
Electric Heaters, 318
Roof Heaters and Pipe Heaters, 320
Overhead Garage Doors, 321
Electric Water Heaters, 322
Electric Motors, 324
Air Conditioners, 326

CHAPTER 14: KNOB-AND-TUBE WIRING 328
Materials Used in Knob-and-Tube Wiring, 328
Spacing the Wires, 330
Spacing the Supports, 330
Installation, 331
Protection against Damage, 334
Three-Wire Installations, 334
Installation of Consuming Devices, 335
Temporary Wiring, 340

CHAPTER 15: MAKING ELECTRICITY A CAREER 342
Industries that Employ Electricians, 342
Careers in Electricity, 344
Information Sources, 347

INDEX 348

CHAPTER 1

Electricity and Electrical Circuits

WHAT IS ELECTRICITY?

Though you cannot see electricity, you are aware of it every day. You see it used in countless ways. You cannot taste or smell electricity, but you can feel it. You can *taste* food cooked with its energy. You can *smell* the gas (ozone) that forms when lightning passes through the air.

Basically there are two kinds of electricity — *static* (stationary) and *current* (moving). This book is chiefly about current electricity. That is the kind of electricity commonly put to use.

Current electricity can be simply defined as the *flow of electrons along a conductor.* To understand that definition, you must know something about chemical elements and atoms.

METRIC MEASUREMENT IN ELECTRICITY

Most of the units used in electricity (such as ampere and volt) are metric units. However, linear and volume measurements given in the National Electrical Code are given in customary measurement. When the SI Metric System is wholly adopted in the United States, the linear and volume measurements used in the Code will probably be stated in metric units. However, such a change is, perhaps, several years away. For this reason, measurements of length and volume in this text have been given in customary units. Conversions to metric have not been made. Table 1-D (Page 41) gives the information necessary to convert customary measurements to metric measurements, and vice versa.

In the metric system of measurement, in numbers of five digits and more, a space (rather than a comma) is inserted every three digits, counting from the decimal point. Thus, 10,000 is written 10 000. The same rule applies to decimals: 0.000001 is written 0.000 001.

ELEMENTS AND ATOMS

Elements are the most basic materials in the universe. Ninety-four elements, such as iron, copper, and nitrogen, have been found in nature. Scientists have made eleven others in laboratories. Every known substance — solid, liquid, or gas — is composed of elements.

It is very rare for an element to exist in a pure state. Nearly always the elements are found in combinations called *compounds*. Even such a common substance as water is a compound, rather than an element. Fig. 1-1.

An *atom* is the smallest particle of an element that retains all the properties of that element. Each element has its own kind of atom. Thus, all hydrogen atoms are alike.

They are different from the atoms of all other elements. However, all atoms have certain things in common. They all have an inner part, the _nucleus_. This is composed of tiny particles called _protons_ and _neutrons_. An atom also has an outer part. It consists of other tiny particles, called _electrons._ The electrons orbit around the nucleus. Figs. 1-2 and 1-3.

Neutrons have no electrical charge, but protons are positively charged. Electrons have a negative charge. Because of these charges, protons and electrons are particles of energy. That is, these charges form an electric field of force within the atom. Stated very simply, these charges are always pulling and pushing each other. This makes energy in the form of movement.

The atoms of each element have a definite number of electrons. They have the same number of protons. For example, a hydrogen atom has one electron and one proton. An aluminum atom has thirteen of each. The opposite charges — negative electrons and positive protons — attract each other and tend to hold electrons in orbit. As long as this arrangement is not changed, an atom is electrically balanced.

However, the electrons of some atoms are easily pushed or pulled out of their orbits. This ability of electrons to move or flow is the basis of current electricity.

Free Electrons

In some materials, heat causes electrons to be forced loose from their atoms. In other materials, such as copper, electrons may be easily forced to drift, even at room temperatures. When electrons leave their orbits, they may move from atom to atom at random, drifting in no particular direction. Electrons that move in such a way are referred to as _free electrons_. However, a force can be applied to direct them in a definite path.

1-1. *Two or more atoms linked are called a molecule. Here are two hydrogen atoms and one oxygen atom which form a molecule of the compound water (H$_2$O).*

TWO HYDROGEN ATOMS AND ONE OXYGEN ATOM = H$_2$O

MOLECULE OF WATER

A. HYDROGEN ATOM

B. ALUMINUM ATOM

NUCLEUS (13 PROTONS, 14 NEUTRONS)
FIRST SHELL
SECOND SHELL
THIRD SHELL
ELECTRON
PROTON

1-2. *Atoms contain protons, neutrons, and electrons.*

Current Flow

If the movement of free electrons is channeled in a given direction, a flow of electrons occurs. This is commonly referred to as *current flow.* Thus, you see that the movement of electrons is related to current electricity.

Energy

Electrons are incredibly small. The diameter of an electron is about 0.000 000 000 000 22 inch. You may wonder how anything so small can be a source of energy. Much of the answer lies in the fact that electrons move at nearly the speed of light. Also, billions of them can move at once through a wire. Speed and concentration produce great energy.

ELECTRICAL MATERIALS

• *Conductors.* A conductor is a material through which electrons move. Actually, all metals and most other materials are conductors to some extent. Some, however, are better than others. Thus, the term *conductor* is usually used to mean a material through which electrons move freely.

What makes one material a better conductor than another? A material that has many free electrons tends to be a good conductor. For practical purposes, however, there are other points that must be considered when choosing a material to use as a conductor.

For example, gold, silver, aluminum, and copper are good conductors. However, the cost of gold and silver limits their use. Copper, because of its superior strength in both hot and cold weather, is preferred over aluminum for many uses.

1-3. *Molecular structure.*

13

• *Insulators.* An insulator is a substance that restricts the flow of electrons. Such materials have a very limited number of free electrons. Thus, you see that the movement of free electrons classifies a material as either a conductor or an insulator. No material is known to be a perfect insulator—that is, entirely void of free electrons. However, there are materials which are such poor conductors that for all practical purposes they are placed in the insulator class.

Wood, glass, mica, and polystyrene are insulators. Fig. 1-4. They have varying degrees of resistance to the movement of their electrons. The higher the line on the chart in Fig. 1-4, the better are the insulating qualities of the material.

• *Semiconductors.* You have heard the word *semiconductor* used in relation to transistors and diodes used in electronic equipment. Materials used in the manufac-

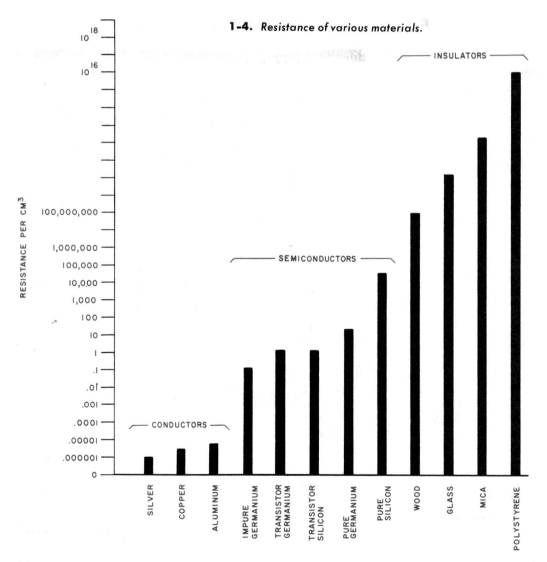

1-4. *Resistance of various materials.*

14

1-5. *The electrostatic precipitator uses a two-stage method of cleansing air by collecting ionized particles on charged plates. Many homes have this type of unit to reduce dust and pollen problems. Clean air is exhausted from the precipitator.*

ture of transistors and diodes have a conductivity between that of a good conductor and a good insulator. Therefore, the name *semiconductor* is given to them. Germanium and silicon are the two most commonly known semiconductor materials. Through the introduction of small amounts of other elements, these nearly pure (99.999 999%) elements become limited conductors. The manufacture of semiconductors is a fascinating process. However, it is too detailed to discuss here. You may wish to research the topic on your own by checking out a book from your library.

GENERATING ELECTRICITY

There are several ways to produce electricity. Remember: *Electricity is the flow of electrons along a conductor.* Friction, pressure, heat, light, chemical action, and magnetism are among the more practical methods used to make electrons move along a conductor. Other methods (sometimes called "exotic") are used to generate electricity for special purposes. For instance, experimental cells developed for the space program are termed "exotic."

• *Friction.* Electricity is produced when two materials are rubbed together. The movement of your shoes against the carpet can cause static electricity. Some practical applications of static electricity are in the manufacture of sandpaper and in the cleaning of polluted air. Fig. 1-5.

COPPER

ELECTRON
FLOW

HOT
JUNCTION

IRON

BURNER

1-6. *A thermocouple.*

• *Pressure.* Electricity is produced when pressure is applied to a crystal. The crystals are usually Rochelle salts or quartz. This is also known as the piezoelectrical effect.

• *Heat.* Electricity is produced when heat is applied to the junction of two dissimilar metals. This junction is usually referred to as a *thermocouple.* The thermocouple is used to measure temperatures in industrial applications. This is especially true in checking the temperature of kilns for ceramic work. Fig. 1-6.

• *Light.* Electricity is produced when light strikes a photosensitive material. (The word *photo* means light). Photoelectric cells are used in cameras, spacecraft, and in radios. Fig. 1-7.

• *Chemical action.* Electricity is produced when a chemical action takes place between two metals in a cell. A single unit is called a cell. Connecting two or more cells together produces a battery. Batteries are used in flashlights, radios, hearing aids, and in calculators. The automobile uses a lead-acid cell combination. You could not start today's cars without a battery. Many types of cells are available today. Fig. 1-8.

• *Magnetism.* Electricity is produced when a magnet is moved past a piece of wire. Or, a piece of wire can be moved through a magnetic field. The result is the same. Motion, a magnetic field, and a piece of wire are needed to produce electricity. To date, magnetism is the most inexpensive way of producing electrical power. We use magnetism to produce electricity for homes.

LIGHT SOURCE

SEMITRANSPARENT
LAYER PASSES
LIGHT AND COLLECTS
PHOTO ELECTRONS

ELECTRON FLOW

1-7. *A photoelectric cell.*

PHOTO SENSITIVE
COPPER OXIDE

PURE COPPER BASE LAYER

1-8. *Charging and discharging cycle of a lead-acid cell.*

• *Exotic generators.* The fuel cell is one of the latest developments for the production of electricity. The *oxygen-concentration cell* includes an electrolyte. The electrolyte conducts an electric charge in the form of oxygen ions, but acts as an insulator to electrons. The electrolyte is located between two electrodes. (The electrolyte is wet. Electrodes are usually metal rods or sheets.) By causing oxygen of different concentrations to pass by the electrodes, it is possible to produce electricity.

1-9. *The MHD generator is an exotic type of power plant.*

The hydrogen-oxygen cell produces water and electricity. Such a cell was used on one of the spaceflights to supply drinking water and electricity in a very small space. Other exotic cells—not all of them perfected yet—are the redox fuel cell, the hydrocarbon fuel cell, the ion-exchange membrane, and the MHD (magnetohydrodynamic) generator. Fig. 1-9. In the MHD generator, hot plasma is generated and seeded in a burner similar to a rocket engine. It then travels through a magnetic field applied at right angles to the flow and past electrodes exposed to this stream of gas. Electrons in the gas are deflected by the field. Between collisions with the particles in the gas, they make their way to one of the electrodes. Electricity flows as the electrons move from the cathode, through the load, to the anode, and back again to the gas stream. There are thousands of other methods of producing electricity. Some of them will now find development funds because of the energy crisis.

VOLTAGE AND CURRENT

So far you have become aware of what electricity is. You have learned some of the ways it is produced. Now it is time to learn how electrical energy is measured. The units of measurement most frequently used are *voltage* and *current.*
- *Volts.* We measure the difference in potential between two plates in a battery in terms of volts. It is actually *electrical pressure* exerted on electrons in a circuit. (A circuit is a pathway for the movement of electrons.) An external force exerted on electrons to make them flow through a conductor is known as *electromotive force,* or EMF. It is measured in volts. Electrical pressure, potential difference, and EMF mean the same thing. The words *voltage drop* and *potential drop* can be interchanged.
- *Current.* For electrons to move in a particular direction, it is necessary for a potential difference to exist between two points of the EMF source. If 6 280 000 000 000 000 000 electrons pass a given point in one second, there is said to be one ampere of current flowing. The symbol for ampere is A. The same number of electrons stored on an object (static charge) and *not moving* is called a *coulomb* (symbol C).

18

Table 1-A. The Greek Alphabet

Name	Capital	Small	Used to designate
alpha	A	α	Angles, area, coefficients, and attentuation constant.
beta	B	β	Angles and coefficients.
gamma	Γ	γ	Electrical conductivity and propagation constant.
delta	Δ	δ	Angles, increment, decrement, and determinants.
epsilon	E	ε	Dielectric constant, permittivity, and base of natural logarithms.
zeta	Z	ζ	Coordinates.
eta	H	$\dot{\eta}$	Efficiency, hysteresis, and coordinates.
theta	Θ	$\vartheta\ \theta$	Angles and angular phase displacement.
iota	I	ι	Coupling coefficient.
kappa	K	κ	
lambda	Λ	λ	Wavelength.
mu	M	μ	Permeability, amplification factor, and prefix *micro*.
nu	N	ν	
xi	Ξ	ξ	
omicron	O	o	
pi	Π	π	Pi = 3.1416.
rho	P	ρ	Restivity and volume charge density.
sigma	Σ	$\sigma\ \varsigma$	Summation.
tau	T	τ	Time constant and time-phase displacement.
upsilon	Υ	υ	
phi	Φ	$\phi\ \varphi$	Magnetic flux and angles.
chi	X	χ	Angles.
psi	Ψ	ψ	Dielectric flux.
omega	Ω	ω	Resistance in ohms and angular velocity.

Current is assumed to flow from negative (−) to positive (+) terminals of a battery or generator.

Current is measured in amperes. In electronics it is sometimes necessary to use smaller units of measurement. The *milliampere* (mA) is used to indicate one one-thousandth of an ampere (0.001 A). If an even smaller unit is needed, it is usually the *microampere* (μA). The microampere is one millionth of an ampere. This may be written as 0.000 001 A. The Greek letter *mu* (μ) is used to indicate *micro*. (Table 1-A lists the Greek alphabet and the terms they designate.)

A voltmeter is used to measure voltage. An ammeter is used to measure current in amperes. A microammeter or a milliammeter may be used to measure smaller units of current.

Resistance

The movement of electrons along a conductor meets with some opposition. This opposition is *resistance*. Resistance is useful in electrical and electronics work. Resistance makes it possible to generate heat, control electron flow, and supply the correct voltage to a device.

Resistance in a conductor depends on four factors: material, length, cross-sectional area, and temperature.

- *Material.* Some materials offer more resistance than others. It depends upon the number of free electrons present in the material.
- *Length.* The longer the wire or conductor, the more resistance it has. The resistance is said to vary directly with the length of the wire.
- *Cross-sectional area.* Resistance varies inversely with the size of the conductor in cross section. In other words, the larger the wire, the smaller the resistance per foot of length.
- *Temperature.* For most materials, the higher the temperature, the higher the resistance. However, there are some exceptions to this in devices known as *thermistors.* Thermistors change resistance with temperature. They decrease in resistance as the temperature increases. Thermistors are used in meters and as temperature indicators.

Resistance is measured by a unit called the *ohm.* The Greek letter *omega* (Ω) is used as the symbol for electrical resistance. The letters of the Greek alphabet and their designations are shown in Table 1-A.

Wire Size

As you become more familiar with electricity and its circuits and with some of the requirements for wiring a house or building, you will become more aware of the current-carrying abilities of wire. The size of the wire is given in numbers. This size usually ranges from 0000 (referred to as *four-ought*) to No. 40. *The larger the wire, the smaller its number.*

For instance, No. 32 wire is smaller than No. 14. Table 1-B shows the resistance (in ohms per 1000 ft [305 m]) in relation to the cross-sectional area. Note how the temperature affects the resistance at 25° C and 65° C [77° F. and 149° F.]. Temperature can make quite a difference in resistance for long wires. Long wires pick up heat and expand when exposed to summer weather.

Copper versus Aluminum Wire

Although silver is the best conductor, its use is limited because of high cost. Two commonly used conductors are aluminum and copper. Each has advantages and disadvantages. For instance, copper has high conductivity and is more ductile (can be drawn out thinner). It is relatively high in tensile strength and can be soldered easily. But it is more expensive than aluminum.

Aluminum has only about 60% of the conductivity of copper. It is used in high-voltage transmission lines and sometimes in home wiring. Its use increased in recent years. However, most electricians will NOT use it to wire a house today. There are several reasons for this. These will become apparent as we progress through the book.

If copper and aluminum are twisted together, as in a wire nut connection, it is possible for moisture to get to the open metals. Corrosion will take place, causing a high-resistance joint. This can result in a dimmer light or a malfunctioning motor.

Table 1-B. Standard Annealed Solid Copper Wire
(American wire gage—B & S)

| Gage number | Diameter (mils) | Cross section | | Ohms per 1000 ft. | | Ohms per mile | Pounds per 1000 ft. |
		Circular mils	Square inches	25°C [77° F.]	65°C [149° F.]	25°C [77° F.]	
0000	460.0	212,000.0	0.166	0.0500	0.0577	0.264	641.0
000	410.0	168,000.0	.132	.0630	.0727	.333	508.0
00	365.0	133,000.0	.105	.0795	.0917	.420	403.0
0	325.0	106,000.0	.0829	.100	.116	.528	319.0
1	289.0	83,700.0	.0657	.126	.146	.665	253.0
2	258.0	66,400.0	.0521	.159	.184	.839	201.0
3	229.0	52,600.0	.0413	.201	.232	1.061	159.0
4	204.0	41,700.0	.0328	.253	.292	1.335	126.0
5	182.0	33,100.0	.0260	.319	.369	1.685	100.0
6	162.0	26,300.0	.0206	.403	.465	2.13	79.5
7	144.0	20,800.0	.0164	.508	.586	2.68	63.0
8	128.0	16,500.0	.0130	.641	.739	3.38	50.0
9	114.0	13,100.0	.0103	.808	.932	4.27	39.6
10	102.0	10,400.0	.00815	1.02	1.18	5.38	31.4
11	91.0	8,230.0	.00647	1.28	1.48	6.75	24.9
12	81.0	6,530.0	.00513	1.62	1.87	8.55	19.8
13	72.0	5,180.0	.00407	2.04	2.36	10.77	15.7
14	64.0	4,110.0	.00323	2.58	2.97	13.62	12.4
15	57.0	3,260.0	.00256	3.25	3.75	17.16	9.86
16	51.0	2,580.0	.00203	4.09	4.73	21.6	7.82
17	45.0	2,050.0	.00161	5.16	5.96	27.2	6.20
18	40.0	1,620.0	.00128	6.51	7.51	34.4	4.92
19	36.0	1,290.0	.00101	8.21	9.48	43.3	3.90
20	32.0	1,020.0	.000802	10.4	11.9	54.9	3.09
21	28.5	810.0	.000636	13.1	15.1	69.1	2.45
22	25.3	642.0	.000505	16.5	19.0	87.1	1.94
23	22.6	509.0	.000400	20.8	24.0	109.8	1.54
24	20.1	404.0	.000317	26.2	30.2	138.3	1.22
25	17.9	320.0	.000252	33.0	38.1	174.1	0.970
26	15.9	254.0	.000200	41.6	48.0	220.0	0.769
27	14.2	202.0	.000158	52.5	60.6	277.0	0.610
28	12.6	160.0	.000126	66.2	76.4	350.0	0.484
29	11.3	127.0	.0000995	83.4	96.3	440.0	0.384
30	10.0	101.0	.0000789	105.0	121.0	554.0	0.304
31	8.9	79.7	.0000626	133.0	153.0	702.0	0.241
32	8.0	63.2	.0000496	167.0	193.0	882.0	0.191
33	7.1	50.1	.0000394	211.0	243.0	1,114.0	0.152
34	6.3	39.8	.0000312	266.0	307.0	1,404.0	0.120
35	5.6	31.5	.0000248	335.0	387.0	1,769.0	0.0954
36	5.0	25.0	.0000196	423.0	488.0	2,230.0	0.0757
37	4.5	19.8	.0000156	533.0	616.0	2,810.0	0.0600
38	4.0	15.7	.0000123	673.0	776.0	3,550.0	0.0476
39	3.5	12.5	.0000098	848.0	979.0	4,480.0	0.0377
40	3.1	9.9	.0000078	1,070.0	1,230.0	5,650.0	0.0299

Circuits

• *Complete circuit.* A complete circuit is necessary for the controlled flow or movement of electrons along a conductor. Fig. 1-10. A complete circuit is made up of a *source of electricity,* a *conductor,* and a *consuming device.* The flow of electrons through the consuming device produces heat, light, or work.

To form a complete circuit, these rules must be followed (refer to Fig. 1-14):

1. Connect one side of the power source to one side of the consuming device (A to B).

2. Connect the other side of the power source to one side of the control device, usually a switch (C to D).

3. Connect the other side of the control device to the consuming device that it is supposed to control (E to F).

This method is used to make a complete path for electrons to flow from one terminal of the battery or power source containing an excess of electrons to the terminal that has a deficiency of electrons. The movement of the electrons along the completed path provides energy. Of course, in order for the path to be complete, the switch must be closed. (Refer to Fig. 1-10.)

If the circuit is so arranged that the electrons have only one path, the circuit is called a *series* circuit. If there are two or more paths for electrons, the circuit is called a *parallel circuit.*

• *Series circuit.* Figure 1-11 shows three resistors connected in series. The current flows through each of them before returning to the positive terminal of the battery.

Kirchoff's Law of Voltages states that the sum of all voltages across resistors or loads is equal to the applied voltage. Voltage drop is considered across the resistor. In Fig. 1-11, the current flows through three resistors. The voltage drop across R_1 is 5 volts. Across R_2 it is 10 volts, and across R_3 it is 15 volts. The sum of the individual voltage drops is equal to the total or applied voltage, 30 volts.

To find the total resistance in a series circuit, just add the individual resistances ($R_T = R_1 + R_2 + R_3 + \ldots$).

• *Parallel circuit.* In a parallel circuit, each load (resistance) is connected directly across the voltage source. There are as many separate paths for current flow as there are branches. Fig. 1-12.

The voltage across all branches of a parallel circuit is the same. This is because all branches are connected across the voltage source. Current in a parallel circuit depends on the resistance of the branch. Ohm's Law (discussed later) can be used to

1-10. *A simple circuit.*

1-11. *A series circuit.*

1-12. *A parallel circuit.*

1-13. *A series-parallel circuit.*

1-14. *An open circuit. Switch in open position.*

determine the current in each branch. You can find the total current for a parallel circuit by *adding the individual currents.* As a formula, this reads:

$$I_T = I_1 + I_2 + I_3 + \ldots$$

The total resistance of a parallel circuit *cannot* be found by adding the resistor values. Two formulas are used for finding parallel resistances. If there are *only two* resistors in parallel, a simple formula can be used:

$$R_T = \frac{R_1 \times R_2}{R_1 + R_2}$$

If there are more than two resistors in parallel, you can use the following formula. (You can also use this formula if there are only two resistors.)

$$\frac{1}{R_T} = \frac{1}{R_1} + \frac{1}{R_2} + \frac{1}{R_3} + \ldots$$

One thing should be kept in mind in parallel resistances: *The total resistance is always less than the smallest resistance.*

• *Series-parallel circuits.* Series-parallel circuits are a combination of the two circuits. Figure 1-13 shows a series-parallel resistance circuit.

• *Open circuit.* An open circuit is one that does not have a complete path for electrons to follow. Such an incomplete path is usually brought about by a loose connection or the opening of a switch. Fig. 1-14.

• *Short circuit.* A short circuit is one that has a path of low resistance to electron flow. It is usually created when a low-resistance wire is placed across a consuming device. The greater number of electrons will flow through the path of least resistance, rather than through the consuming device. A short usually generates an excess current flow. This results in overheating, possibly causing a fire or other damage. Fig. 1-15.

It is easy to compute the amount of current flowing in a circuit if the voltage and the resistance are known. The relationship between voltage, current, and resistance in any circuit is shown by *Ohm's Law.*

OHM'S LAW

There are three basic quantities of electricity, and each has a relationship to the other two. A physicist named Georg S. Ohm discovered the relationship in 1827. He found that *in any circuit where the only opposition to the flow of electrons is resistance, there is a relationship between the values of voltage, current, and resistance.* The strength or intensity of the current is directly proportional to the voltage and inversely proportional to the resistance.

It is easier to work with Ohm's Law when it is expressed in a formula. In the formula, **E** represents EMF, or voltage. **I** is the current, or the intensity of electron flow. **R** stands for resistance. The formula is $E = I \times R$. This is the formula to use to find the EMF (voltage) when the current and resistance are known.

To find the current when the voltage and resistance are known, the formula to use is:

$$I = \frac{E}{R}$$

To find the resistance when the voltage and current are known, the formula to use is:

$$R = \frac{E}{I}$$

Using Ohm's Law

There are many times in electrical work when you will need to know Ohm's Law. For

1-15. *A short circuit. The lamp will not glow.*

1-16. *Ohm's Law. Place a finger on the unknown value. The remaining two letters will give the formula to use for finding the unknown value.*

example, you need to know Ohm's Law to determine wire size in a particular circuit or to find the resistance in a circuit.

The best way to become accustomed to solving problems is to start with a simple problem such as the following:

1. If the voltage is given as 100 volts and the resistance is known to be 50 ohms, it is a simple problem and a practical application of Ohm's Law to find the current in the circuit.

$$I = \frac{E}{R}$$

$$I = \frac{100 \text{ volts}}{50 \text{ ohms}}$$

$$I = 2 \text{ amperes}$$

2. If the current is given as 4 amperes (shown on an ammeter) and the voltage (read from a voltmeter) is 100 volts, it is easy to find the resistance.

$$R = \frac{E}{I}$$

$$R = \frac{100 \text{ volts}}{4 \text{ amperes}}$$

$$R = 25 \text{ ohms}$$

3. If the current is known to be 5 amperes, and the resistance is measured (before current is applied to the circuit) and found to be 75 ohms, it is then possible to determine how much voltage is needed to cause the circuit to function properly.

$$E = I \times R$$

$$E = 5 \text{ amperes} \times 75 \text{ ohms}$$

$$E = 375 \text{ volts}$$

Figure 1-16 illustrates the way the formula works.

POWER

Power is defined as the rate at which work is done. It is expressed in metric measurement terms of *watts* (for power) and in *joules* (for energy or work). A watt is the power that gives rise to the production of energy at the rate of one joule per second (W = J/s). A joule is the work done when the point of application of a force of one newton is displaced a distance of one meter in the direction of the force. (J = N × m).

It has long been the practice in the United States to measure work in terms of *horsepower.* Electric motors are still rated in horsepower and probably will be for some time.

Power can be electrical or mechanical. When a mechanical force is used to lift a weight, work is done. The rate at which the weight is moved is called power. Horsepower is defined in terms of moving a certain weight over a certain distance in one minute. Energy is consumed in moving a weight, or work is done. The findings in this field have

Table 1-C. Horsepower

One horsepower is usually defined as the amount of work required to move a 550-pound weight a distance of one foot in one second.

In most cases the *modern* way to measure power is in kilowatts rather than horsepower. In case a motor is specified in terms of horsepower, but is rated in watts or kilowatts, the conversion is simple:

1 horsepower = 746 watts

Divide the number of watts or kilowatts by 746 or 0.746, respectively, to find the horsepower rating.

been equated with the same amount of work done by electrical energy. *It takes 746 watts of electrical power to equal one horsepower.* Table 1-C.

However, it should be remembered that a one-horsepower motor uses about 1 100 watts to produce the power. The extra watts are consumed in heat and friction losses.

The horsepower rating of electric motors is obtained by taking the voltage and multiplying it by the current drawn under full load. This power is measured in watts. In other words, one volt times one ampere equals one watt. When put into a formula:

Power = Volts × Amperes, or P = E × I

Kilowatts

The prefix *kilo* means one thousand. Thus, one thousand watts equal one kilowatt. The abbreviation for kilowatt is kW. There is also a unit known as the kilowatt-hour. It is abbreviated kWh and is equivalent to one thousand watts used for one hour. Electric bills are figured in kilowatt-hours. Usage for an entire month is computed on an hourly basis and then read in the kWh unit.

Power formulas are sometimes needed to figure the wattage of a circuit. Here are the three most commonly used formulas:

$$P = E \times I$$

$$P = \frac{E^2}{R}$$

$$P = I^2 \times R$$

This means that if any one of the three — voltage, current, or resistance — is missing, it is possible to find the missing quantity by using the relationship of the two known quantities. However, you must also know the power consumed. In later chapters you will encounter the problem of the I^2R *losses* and some other terminology related to the formulas just shown.

The milliwatt (mW) is sometimes used in referring to electrical equipment. For instance, the rating of the speaker in a transistor radio may be given as 100 mW or 300 mW. This means a 0.1 watt or 0.3 watt rating, since the prefix *milli* means one-thousandth. Transistor circuits are designated in milliwatts, but power-line electrical power is usually in kilowatts.

MEASURING ELECTRICITY

Electricity must be measured if it is to be sold, or if it is to be fully utilized. There are a number of ways to measure electricity. It can be measured in volts, amperes, or watts. The kilowatt-hour meter is the device most commonly used to measure power.

Meters

To measure anything, there must be a basic unit in which to measure. In electricity, the current (flow of electrons) is measured in a basic unit called the ampere. The current is usually measured with a permanent magnet and an electromagnet arranged to indicate the amperes. Such a device is necessary since we are unable to see an electron—even with the most powerful microscopes. Obviously, counting the number of electrons passing a given point in a second is impossible when there are no visible particles to count. Therefore, a magnetic field is used to measure the effect of the electrons.

The D'Arsonval meter movement uses a permanent magnet as a base over which a wire or electromagnet is pivoted and allowed to move freely. When current flows through the coil, a magnetic field is set up. Figs. 1-17a and 1-17b. The strength of the magnetic field determines how far the coil will be deflected. The polarity of the moving coil is the same as that of the permanent magnet. A repelling action results. This is in proportion to the strength of the magnetic field generated by the current flowing through the coil. The number of turns in the coil, times the current through the coil, determines the strength of the magnetic field. Since the meter coil is pivoted on jeweled bearings to reduce any friction, the movement is calibrated against a known source of current or against another meter. The scale on a new device is calibrated to read in amperes, milliamperes, or microamperes. Fig. 1-17c.

1-17A. *D'Arsonval meter movement.*

1-17B. *D'Arsonval meter movement showing completed unit in diagram form.*

POINTER

JEWEL BEARING

COIL

POINTER

HORSESHOE MAGNET

UPPER HAIRSPRING

POLE PIECE

IRON CORE

SUPPORT FOR CORE

LOWER HAIRSPRING

ASSEMBLED ARRANGEMENT

1-17C. *Assembled D'Arsonval meter movement.*

METER

1 mA M.M

Resistor

1 mA

9 mA

SHUNT

10 mA

10 mA

EXTERNAL CIRCUIT

1-19. *A meter movement with a shunt to increase its range to 10 mA.*

1-20. *A multimeter using a D'Arsonval meter movement. (EICO)*

1-18. *An AC meter movement made by adding a diode to a DC meter movement.*

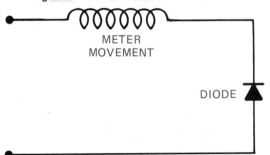

METER MOVEMENT

DIODE

Diode ↙

- *AC ammeter.* If alternating current is to be measured by a DC meter movement, a rectifier is inserted in the circuit (meter circuit) to change the AC to DC. It can then be measured by the meter movement. Otherwise the alternating current will make the needle on the meter vibrate rapidly. This vibration means there is little or no movement from zero. Fig. 1-18.
- *Shunts.* A shunt is a means of bypassing current around a meter movement. A resistor of the proper size is inserted *across* the meter movement to bypass the current around the movement. Most of the current is bypassed. Only the necessary amount is left to cause the meter to deflect at its designed limit. The meter is calibrated to read on its scale the full amount of current flowing in the circuit. Fig. 1-19.

A number of shunts can be placed in a meter case and switched. A different resistor (shunt) can be switched in for each range needed. A meter with more than one range is called a *multimeter.*

A multimeter (one which can measure volts and ohms as well) is shown in Fig. 1-20. It can measure from 0 to 1 milliampere, 0 to 10 milliampere, and from 0.1 to 1 ampere. This is a 1 milliampere meter movement. It needs no shunt when used to measure as high as 1 milliampere. A shunt is switched in, however, to measure a range of 0 to 10 milliampere and 0.1 to 1 ampere. A switch on the meter can also add a <u>diode</u> to help with measuring AC.

Common and *positive* holes have test leads inserted to attach to the circuit being measured. Common is negative (−) or black, and the positive is (+) or red. However, polarity isn't necessary to measure AC. Either lead can be used at any terminal in an AC circuit. *Ammeters are always placed in series in a circuit.* This usually means the circuit has to be broken and the meter inserted in the line.

- *Voltmeter.* The voltmeter measures electrical pressure, or volts. It is nothing more than an ammeter with a resistor added in the meter circuit. The high resistance of the voltmeter makes it possible to place it across a power source (in parallel). Fig. 1-21. A number of resistors, called *multipliers,* can be switched into a meter circuit to increase its range or make it capable of measuring higher voltages. The voltmeter in Fig. 1-22 is capable of measuring from 0 to 150, 0 to 300, and 0 to 750 volts by placing

1-21. *A voltmeter's internal circuit. The multiplier extends the meter movement's range.*

INPUT

IN

MULTIPLIER

Resistor ↗

1-22. *DC voltmeter. (Weston)*

the proper multiplier into the meter circuit. Note how the terminals on top of the meter allow for placing test probes in a number of different positions. This allows the measurement range to be varied.

• *Ohmmeter.* The basic unit for measuring resistance is the *ohm.* An ohmmeter is a device used to measure resistance or ohms (Ω). It is an ammeter (or milliammeter or micrometer) movement, modified to measure resistance. Fig. 1-23.

Figure 1-23 shows a multimeter capable of measuring ohms with three different ranges: R × 1, R × 100, and R × 10K. This means it can measure from 0 to 200 ohms on the R × 1 scale and 0 to 200 000 ohms on the R × 100 range. Within the R × 10K or R × 10 000 range, it is capable of measuring from 0 to 20 000 000 ohms or 0 to 20 megohms (mega means one million). The meter scale has to be multiplied by the 100 or 10 000 number to have it read the proper value. By changing resistors, it is possible to vary the resistance measuring range of an ohmmeter. Figure 1-24 shows a basic ohmmeter using a 1-milliampere meter movement and its necessary parts. Note how the battery serves as the power source. This makes it necessary to turn off power whenever you read the resistance of a circuit. THE OHMMETER HAS ITS OWN POWER SOURCE. DO NOT CONNECT IT TO A LIVE CIRCUIT OR ONE WITH THE POWER ON. To do so will result in the destruction of the meter movement. Fig. 1-25.

Adjust the ohmmeter so that the meter reads zero before starting to use it to measure resistance. Fig. 1-26. This means you have adjusted the meter circuit to compensate for the battery voltage changes. Battery voltages decrease with shelf life. It doesn't matter whether or not the battery is used. It will, in time, lose its voltage.

• *Clamp-on meters.* Figures 1-27 and 1-28 show two different styles of an AC clamp-on meter. They are inserted over a wire carrying alternating current. The magnetic field around the wire induces a small amount of current in the meters. The scale

1-23. *A multimeter. This type is used to measure ohms, volts, and amperes. (Heath)*

1-24. *The basic circuit of an ohmmeter.*

1 mA
M.M.

OHMS
ADJUST

CURRENT
LIMITING

DC
POWER

TEST LEADS

MEASURED

RESISTOR

1-25. *Test leads are touching the leads of a resistor to measure its resistance.*

SHORTED

1-26. *Test leads are crossed (shorted) so resistor can be adjusted to make the meter read zero.*

is calibrated to read amperes or volts. Because the wire is run through the large loop extending past the meter movement, it is possible to read the AC voltage, or current, without removing the insulation from the wire. These meters are very useful when working with AC motors.

• *Wattmeter.* A wattmeter measures electrical power. Electrical power is figured by multiplying the voltage times the current. A wattmeter has electromagnetic coils (a coil with many turns of fine wire for voltage and a coil with a few turns of heavy wire for

$$E/p = V \cdot C \qquad Ep = E \cdot I$$

1-27. *A clamp-on type of portable AC voltmeter-ammeter. (Weston)*

1-28. *A clamp-on type of AC volt-ammeter. (Weston)*

current). The voltage coil is connected across the incoming line. The current coil is inserted in series with one of the incoming wires. Two coils are stationary and in series with a moving coil. The strength of the magnetic fields determines how much the moving coil is deflected. The deflection of the needle is read on a scale calibrated in watts. In this way the wattmeter measures the power consumed in one second.

- *Kilowatt-hour meter.* The kilowatt-hour meter was designed for measuring the electrical power used over a long period of time. The kWh meter is often seen on the side of a house or other building. It measures power used over a certain time period, such as a month. A kilowatt-hour meter measures the power consumed in terms of thousands of watts. Electric power is priced at a certain rate per kWh.

The kilowatt-hour meter is a small induction motor. Meter torque is produced by an electromagnet called a stator, which has two sets of windings. One winding, called a *potential coil*, produces a magnetic field representing circuit voltage. Another winding, known as a *current coil*, produces a magnetic field that represents the load current. These two coils are arranged so that their magnetic fields create a force on the meter disc. This force is directly proportionate to the power, or watts, drawn by the connected load.

Permanent magnets are used to introduce a retarding, or braking force, which is proportionate to the speed of the disc. The magnetic strength of these retarding magnets regulates the disc speed for any given load so that each revolution of the disc always measures the same quantity of energy or watt-hours. Disc revolutions are converted to kilowatt-hours on the meter register.

Most meters are inserted into a socket on the wall of a structure. Removing the meter interrupts or terminates power without handling of dangerous high voltage wires. Kilowatt-hour meters are tested by computers in the service centers of power companies. Fig. 1-29.

Other Types of Meters

There are other types of meters used to measure voltage and current. The D'Arsonval movement is only one of many types used today. The *taut band* type is basically the same as the D'Arsonval. However, a tightly stretched and twisted band is used to hold the coil and needle in place between the permanent magnet poles. In addition, no moving points touch the meter case. Thus, jeweled bearings are unnecessary. The band is twisted when it is inserted into the meter frame. Thus, it will cause the coil to spring back to its original resting place upon interruption of current through the coil.

- *Electrodynamometer.* The electrodynamometer type of meter uses no permanent magnet. Two fixed coils produce the magnetic field. The meter also uses two moving coils. This meter can be used as a voltmeter or an ammeter. It is not as sensitive as the D'Arsonval meter movement.

- *Inclined-coil iron-vane meter.* The inclined-coil iron-vane meter is used for measuring AC or DC where large amounts of current are present. This meter can be used as a voltmeter or ammeter.

CONTROLLING ELECTRICITY

To make electricity useful it is necessary to control it. You want it in the proper place at the proper time. Otherwise, it can do great damage — even kill. Electricity can

1-29. *A kilowatt-hour meter.*

be controlled by using switches, relays, or diodes. These devices are used to direct the current to the place where it will work for you. Each device is carefully chosen to do a specific job. For example, the relay is used for remote-control work. A diode is used to control large and small amounts of current in electrical as well as electronic equipment. A *diode* is a device which allows current to flow in one direction only. It can be used to change AC to DC.

Switches

There are a number of switches used for controlling electricity. Each switch has a different name. This helps designate it according to the job it performs. For instance, the single-pole, single-throw switch (SPST) is just that. It is a single pole that is moved either to *make connection* between two points or to *not make connection*. In the off position the contacts are not touching, and the flow of electrons is interrupted.

The double-pole, double-throw switch (DPDT) can be used to control more than one circuit at the same time. It can be used to reverse the direction of rotation of a DC motor by reversing polarity. *such as 220 Volts*

The double-pole, single-throw switch (DPST) is used to control two circuits at the same time. It can be used as a simple on-off switch for two circuits. When it is open it interrupts the current in the two circuits. When it is closed it completes the circuits for proper operation. This, too, is an open switch. It is meant to be used only on low voltages where the danger of shock is very much reduced.

The doorbell switch or door chime switch is extremely simple. It completes a circuit from the low-voltage transformer to the chime or bell. Fig. 1-30. When the button is pressed, it completes the circuit to the bell or chime, causing current to flow from the transformer to the chime. Fig. 1-31.

1-30. *A door bell or chime circuit. Switch open.*

1-31. *A door bell or chime circuit. Switch closed.*

• *Toggle switches.* Toggle switches are used to turn various devices on and off, or to switch from one device to another. They are made in a number of configurations to aid in selection for a particular job.

• *Residential toggle switches.* Various shapes are encountered when switches are needed for use in business, industry, or the home. For instance, in Fig. 1-32 you will find examples of some of those used to switch 120 volts AC and 240 volts AC in common circuits used in lighting and small motors. In Fig. 1-32, switch A is a residential toggle switch, rated at 10 amperes on a 125-volt line. Note the absence of "plaster ears" near the long screws. Switch B is like switch A, except that it has wide plaster ears. These can be removed easily if not required to hold the switch rigid in its box. The plaster ears are scored (marked) so that they can be bent easily and removed. Switch C is more expensive than either A or B. It has *specification grade* (top of the line) quality with wide plaster ears. Note that the screws for attaching circuit wiring are located topside, instead of along the side of the switch.

Switch D is a high-capacity, heavy-duty, industrial type, rated for 20 amperes at 125 volts. Switch E is an extra heavy-duty industrial type. It is a more expensive switch that minimizes the arcing of contacts when turned on and off. The arc, which occurs each time a switch is turned on or off, creates high heat. The heat can cause the con-

1-32. *Switches used for residential wiring.*

| A | B | C | D | E | F | G | H |

tacts of the switch to become pitted and make a high-resistance contact. The contacts in switch E have an extended life, made possible by the use of arc snuffers.

Switch F is a "no-klik," or quiet, heavy-duty switch. It has eliminated the noise associated with the on-off operation of a switch. Switch G is also heavy-duty, quiet, of high capacity, specification grade, and good for 15 amperes at 120 or 277 volts. Switch H is a quiet 15-ampere AC, side- or back-wired with binding screw, and pressure, or screwless, terminals. Some switches have a *grounding strap* designed for use with nonmetallic systems that use Bakelite boxes and for bonding between the device strap and steel boxes. These newer switches feature a green grounding-screw terminal. They are available in either ivory or brown.

• *Three-way switches.* Three-way switches are used where you need to control a light or device from more than one location. These switches have three terminals instead of two. They do not have the words *on* and *off* on the handles. More discussion of this type of switch and its circuitry will follow in a later chapter. The three-way switch looks like a regular switch with the exception of the three terminals for wiring into a circuit.

• *Four-way switches.* Four-way switches are double pole. They are used where a light or device needs to be controlled from three or more locations. If three controls are preferred, you need two three-way switches and one four-way switch. The four-way resembles the three-way, but has four terminals for connection into a circuit. It does not have *on* or *off* on its handle, since either up or down may be on or off, as you will see later. More information concerning the four-way switch will be given in a later chapter.

Other types of switches are available for different jobs of current control. They will be shown later, as they are introduced in connection with a specific job.

Switches are used to turn the flow of electricity on or off, thereby causing a device to operate or cease operation. Switches can be used to reverse polarity and, as in the case of electric motors, the direction of rotation can be reversed by this action. Switches, as you have already seen, come in many shapes and sizes. The important thing to remember is to use a switch with proper voltage and current rating for the job to be done. A careful study of the types presented in this chapter will help in the proper selection of a switch for a particular job.

Solenoids

Solenoids are devices that turn electricity, gas, oil, or water on and off. Solenoids can be used, for example, to turn the cold water on, and the hot water off, to get a proper mix of warm water in a washing machine. To control the hot water solenoid, a thermostat is inserted in the circuit.

Figure 1-33 shows a solenoid for controlling natural-gas flow in a hot-air furnace. Note how the coil is wound around the plunger. The plunger is the core of the solenoid. It has a tendency to be sucked into the coil whenever the coil is energized by current flowing through it. The electro-magnetic effect causes the plunger to be attracted upward into the coil area. When the plunger is moved upward by the pull of the electromagnet, the soft disc (10) is also pulled upward, allowing gas to flow through the valve. This basic technique is used to control water, oil, gasoline, or any other liquid or gas.

The starter solenoid on an automobile uses a similar procedure. However, the plunger has electrical contacts on the end that complete the circuit from the battery to the

1. COVER NUT
2. CONDUIT OPENING
3. COIL
4. PLUNGER
5. BODY UNION NUT
6. CENTERING WASHER
7. LIFT WASHER
8. VALVE STEM
9. VALVE BODY
10. SOFT DISC
11. VALVE SEAT
12. PLUNGER SPRING

1-33. *A solenoid for controlling natural gas flow to a hot-air home furnace. (Honeywell).*

starter. The solenoid uses low voltage (12 volts) and low current to energize the coil. The coil in turn sucks the plunger upward. The plunger then touches heavy-duty contacts that are designed to handle the 300 amperes needed to start a cold engine. In this way, low voltage and low current are used, from a remote location, to control low voltage and high current.

Solenoids are *electromagnets.* An electromagnet is composed of a coil of wire wound around a core of soft iron. When current flows through the coil, the core will become magnetized.

The magnetized core can be used to attract an armature and act as a magnetic circuit breaker. Fig. 1-34. (A circuit breaker, like a fuse, protects a circuit against short circuits and overloads.) In Fig. 1-34, the magnetic circuit breaker is connected in series with both the load circuit to be protected and with the switch contact points. When excessive current flows in the circuit, a strong magnetic field in the electromagnet causes the armature to be attracted to the core. A spring attached to the armatures causes the switch contacts to open and break the circuit. The circuit breaker must be reset by hand to allow the circuit to operate properly again. If the overload is still present, the circuit breaker will "trip" again. It will continue to do so until the cause of the short circuit or overload is found and corrected.

Relays

A relay is a device that can control current from a remote position through use of a separate circuit for its own power. Figure 1-35 shows a simple relay circuit.

When a switch is closed, current flows through the electromagnet, or coil, and energizes it. The pull of the electromagnet causes the soft iron armature to be attracted toward the electromagnet core. As the armature moves toward the coil, it touches the contacts of the other circuits, thereby completing the circuit for the load. When a switch opens, the relay coil *de-energizes,* and the spring pulls the armature back. This action breaks the contact and removes the load from the 12-volt battery. Relays are remote switches that can be controlled from almost any distance if the coil is properly wired to its power source.

Many types of relays are available. They are used in telephone circuits and in almost all automated electrical machinery.

RESISTORS

A resistor is a device used to provide a definite, required amount of opposition to current in a circuit. Resistance is the basis for the generation of heat. It is used in circuits to control the flow of electrons and to insure that the proper voltage reaches a particular device.

Resistors are usually classified as either wire-wound or carbon-composition. The symbol for a resistor of either type is —⋀⋀⋀— .

1-34. *A magnetic circuit breaker.*

SPRINGS

ARMATURE

LOAD CIRCUIT
IN SERIES WITH COIL

POWER SUPPLY

ELECTROMAGNET   LOAD

CONTACT

SOFT IRON
ARMATURE

SPRING

SWIVEL PIN

12
VOLTS

1-35. *A simple relay circuit.*

SWITCH

- *Wire-wound resistors* are used to provide sufficient opposition to current flow to dissipate power of five watts or more. They are made of resistance wire. Fig. 1-36. Variable wire-wound resistors are also available for use in circuits where voltage is changed at various times. Fig. 1-37. Some variable resistors have the ability to be varied but also adjusted for a particular setting. Fig. 1-38.

High wattage, wire-wound resistors may be purchased in many sizes and shapes.

- *Carbon-composition* resistors are usually found in electronic circuits of low wattage, since they are not made in sizes beyond 2 watts. They can be identified by three- or four-color bands around them. Their resistance can be determined by reading the color bands and checking the resistor color code.

RESISTOR COLOR CODE

| 0 Black | 2 Red | 4 Yellow | 6 Blue | 8 Gray |
| 1 Brown | 3 Orange | 5 Green | 7 Violet | 9 White |

1-36. A wire-wound resistor. (Ohmite)

1-37. A variable wire-wound resistor. Note it has three terminals for connections. (Ohmite)

1-38. An adjustable wire-wound resistor. (Ohmite)

¼ WATT

.390

½ WATT

.416

1 WATT

.593

2 WATT

.728

1-39. *Carbon composition resistors.*

1-40. *A 47 000-ohm resistor.*

YELLOW
VIOLET
ORANGE ← *if gold or silver you divide*
SILVER

47 000 OHMS

1-41. *A 2.4-ohm resistor.*

RED
YELLOW } 24
GOLD ——— = 2.4
SILVER 10

2.4 ohms ± 10%

2.4 OHM ± 10%

The wattage ratings of carbon-composition resistors are determined by physical size. They come in ¼-watt, ½-watt, 1-watt, and 2-watt sizes. By examining them and becoming familiar with them through use, you should be able to identify wattage rating by sight. The larger the physical size of the resistor, the larger the wattage rating. Fig. 1-39.

Resistor Color Code

Take a close look at a carbon-composition resistor. The bands should be to your left. Fig. 1-40. Read from left to right. The first band gives the first number according to the color code. In this case it is yellow, or 4. The second band gives the next number, which is violet, or 7. The third band represents the *multiplier* or *divisor.* If the third band is a color in the 0 to 9 range in the color code, it states the number of zeros to be added to the first two numbers. Orange is 3. Thus, the resistor in Fig. 1-40 has a value of 47 000 ohms resistance.

If there is no fourth band, the resistor has a tolerance rating of ±20%. (± means "plus or minus.") If the fourth band is silver, the resistor has a tolerance of ±10%. If the fourth band is gold, the tolerance is ±5%.

Silver or gold may also be used for the third band. In this case, according to the color code, the first two numbers (obtained from the first two color bands) must be *divided* by 10 or 100. Silver means divide by 100; gold means divide by 10. For example, if the bands on a resistor are red, yellow, gold, and silver, the resistance would be 24 divided by 10, or 2.4 ohms ±10%. Fig. 1-41.

39

Table 1-D. Conversions

To convert-	Into-	Multiply by-	Conversely multiply by-
Inches	Centimetres	2.54	0.3937
Inches	Mils	1000	0.001
Joules	Foot-pounds	0.7376	1.356
Joules	Ergs	10^7	10^{-7}
Kilogram-calories	Kilojoules	4.186	0.2389
Kilograms	Pounds (avoirdupois)	2.205	0.4536
Kg per sq metre	Pounds per sq foot	0.2048	4.882
Kilometres	Feet	3281	3.048×10^{-4}
Kilowatt-hours	Btu	3413	2.93×10^{-4}
Kilowatt-hours	Foot-pounds	2.655×10^6	3.766×10^{-7}
Kilowatt-hours	Joules	3.6×10^6	2.778×10^{-7}
Kilowatt-hours	Kilogram-calories	860	1.163×10^{-3}
Kilowatt-hours	Kilogram-metres	3.671×10^5	2.724×10^{-6}
Litres	Cubic metres	0.001	1000
Litres	Cubic inches	61.02	1.639×10^{-2}
Litres	Gallons (liq US)	0.2642	3.785
Litres	Pints (liq US)	2.113	0.4732
Metres	Yards	1.094	0.9144
Metres per min	Feet per min	3.281	0.3048
Metres per min	Kilometres per hr	0.06	16.67
Miles (nautical)	Kilometres	1.853	0.5396
Miles (statute)	Kilometres	1.609	0.6214
Miles per hr	Kilometres per min	2.682×10^{-2}	37.28
Miles per hr	Feet per min	88	1.136×10^{-2}
Miles per hr	Kilometres per hr	1.609	0.6214
Poundals	Dynes	1.383×10^4	7.233×10^{-5}
Poundals	Pounds (avoirdupois)	3.108×10^{-2}	32.17
Sq inches	Circular mils	1.273×10^6	7.854×10^{-7}
Sq inches	Sq centimetres	6.452	0.155
Sq feet	Sq metres	9.29×10^{-2}	10.76
Sq miles	Sq yards	3.098×10^6	3.228×10^{-7}
Sq miles	Sq kilometres	2.59	0.3861
Sq millimetres	Circular mils	1973	5.067×10^{-4}
Tons, short (avoir 2000 lb)	Tonnes (1000 kg)	0.9072	1.102
Tons, long (avoir 2240 lb)	Tonnes (1000 kg)	1.016	0.9842
Tons, long (avoir 2240 lb)	Tons, short (avoir 2000 lb)	1.120	0.8929
Watts	Btu per min	5.689×10^{-2}	17.58
Watts	Ergs per sec	10^7	10^{-7}
Watts	Ft-lb per min	44.26	2.26×10^{-2}
Watts	Horsepower (550 ft-lb per sec)	1.341×10^{-3}	745.7

Table 1-D. Conversions (Continued)

To convert-	Into-	Multiply by-	Conversely multiply by-
Watts	Horsepower (metric) (542.5 ft-lb per sec)	1.36×10^{-3}	735.5
Watts	Kg-calories per min	1.433×10^{-2}	69.77

<table>
<tr><td colspan="2" align="center">Symbols</td></tr>
<tr><td>hertz</td><td>Hz</td></tr>
<tr><td>kilohertz</td><td>kHz</td></tr>
<tr><td>megahertz</td><td>MHz</td></tr>
<tr><td>RPM, rpm</td><td>r/min</td></tr>
</table>

Conversions

1 mile	= 1.609 344 km
1 cubic foot (cu. ft.)	= 0.028 316 846 592 cubic metres (m^3)
1 ton (short)	= 0.907 184 74 metric ton
1 foot	= 0.304 8 metre

Cycles per second, cps or ~, has been changed to *hertz* per second. The time interval of the *second* is understood to be there. So, it is referred to simply as hertz without the *per second* added. Instead of 60 cps, 60 cycles per second, or 60 ~, it is now 60 hertz or 60 Hz.

Revolutions per minute, rpm, or RPM, has been changed to r/min. This will probably be showing up later as the switch to metric becomes more noticeable.

Resistors are available in hundreds of sizes and shapes. Once familiar with electronics and electrical circuits, you will be able to identify various components by their shape, size, or markings. Products for such circuits are constantly changing. New items are being marketed almost every day. To stay informed about these products, it is necessary to read the literature written about the industry. Each area of electrical energy has its own magazines to keep those on the job informed and up-to-date in their special fields of interest.

QUESTIONS

1. What is electricity?
2. What is an atom?
3. What is an element?
4. What is the difference between static electricity and current electricity?
5. What is a conductor?
6. What is an insulator?
7. What is an exotic generator?
8. What is the unit used to measure voltage?
9. What is the unit used to measure current? *amps*
10. What is the unit used to measure resistance? *ohms*
11. What is the relationship of wire size to the numbering system used to designate its diameter?
12. What is a complete circuit?
13. What is an open circuit?
14. What is a short circuit?
15. State Ohm's Law.
16. What is a kilowatt-hour?
17. How many watts are there in one horsepower?
18. What is the purpose of a meter shunt?
19. How can you convert a DC meter movement to measure AC voltage?
20. What is the difference between a three-way switch and a four-way switch?
21. What is the definition of a solenoid?
22. What is a resistor?
23. What is the value of a resistor which has color bands that read: red, orange and yellow?

CHAPTER 2

Producing Electricity

As discussed in Chapter 1, there are a number of ways to produce electricity, some of which are commercially feasible. The use of magnetism is the most common method of generating electricity in large quantities.

Cells, or batteries, produce direct current (DC). A more economical way of producing DC, however, is with a mechanically driven generator. Mechanical force is used to rotate a wire loop in a magnetic field to generate electricity. The magnetic field is generated by a current-carrying wire, looped around a core. Fig. 2-1.

Since the current from the battery supply is DC with a negative (−) and a positive (+) polarity, the north (N) and south (S) fields are fixed in the positions indicated. If a conductor is moved in an upward direction through the magnetic field between the N and S poles (Fig. 2-1, A), a current will flow in the conductor in the direction shown by the arrows. But if the conductor moves downward (as in Fig. 2-1, B), then current flows in the opposite direction. The voltage generated will depend upon the following: the intensity of the magnetic field, the number of turns of the wire, and the speed at which the wire passes through the magnetic field.

2-1. *The magnetism principle of electrical generation.*

FIELD POLES

VOLTAGE

TIME

VOLTAGE ACROSS LOOP

BRUSH

S

N

ARMATURE

COMMUTATOR

VOLTAGE

TIME

VOLTAGE ACROSS BRUSHES

2-2. *A simplified DC generator with output voltages shown.*

A simple generator is shown in Fig. 2-2. A loop of wire is wrapped around an iron core called an *armature.* Copper segments form the *commutator.* These segments are attached to the ends of the loop of wire. They are insulated from the core and from each other. *Brushes* (conductors that make sliding contact) are so placed that they contact the commutator. They carry any generated electricity to the load, or consuming device. To produce electricity the armature must be mounted between *field coils.* Thus, the magnetic force generated by the electromagnet will be cut by the rotating armature. Field coils form an electromagnet when they are wrapped around soft-iron cores, known as *field poles.*

The position of the armature in Fig. 2-2 represents the point at which the armature loop cuts directly across the magnetic field. At this point it is generating its maximum output. When the armature has rotated another one-quarter turn, it will be moving parallel with the magnetic field. No output will be obtained from the generator. During the time that the armature rotates 360° (one revolution), it generates a maximum and a minimum twice. It generates a maximum when passing across the S pole, and again when crossing the N pole.

Voltage across the loop is represented as alternating current (AC) in Fig. 2-2. Voltage across the brushes is shown as pulsating direct current (PDC). The commutator acts as a reversing switch as the armature rotates in the different fields. As a result of the switching action, the current output is a series of maximums and minimums with current flowing in only one direction.

● *AC generator.* An AC generator is usually referred to as an *alternator.* Alternators generate most of the electrical power used today. Large generating plants produce the electricity demanded by a world constantly increasing its need for energy.

Alternating current (AC) generators operate on the basic principle illustrated in Fig. 2-3. Atomic- and steam-powered generators use the same basic idea of electric power generation.

POWER PLANT OPERATIONS – AC

An alternating current (battery-less) plant supplies electricity directly to the line to which electrical consuming devices are connected. It must be operated whenever any part, or all of the equipment, is to be used.

This type of plant supplies primary, or portable, power for field or construction work. It will operate electric lights, hand tools with universal motors, and other equipment and motors designed for AC. Alternating current power has numerous other uses, especially for the home, farm, factory, industry, hospital, and communications fields. Its vital function in these fields is that of a reliable source of *emergency* electricity when there is an interruption in regular power service. Fig. 2-4.

Much of the equipment and most of the motors built today require alternating current. Thus, the AC plant is more widely used than any other type of generating plant.

An AC plant has other advantages. It is available in the widest range of sizes and with more types of controls (ranging from manual-start to full automatic). It will also operate a greater selection of low-cost equipment. Many appliances and some lighting, such as fluorescent, can be operated only on AC. Wiring distribution of the alternating current system is probably the most flexible, efficient, and least expensive to operate.

The AC plant is ideal as a primary source of power. It is capable of operating a variety of large-capacity equipment on regular high-line power (overhead wires) when a

2-3. *A simplified alternator, or AC generator.*

LIGHT
(POWER ABSORPTION)

STATOR

ROTOR

Ⓐ

SLIP RINGS

Ⓑ

BRUSHES

2-4. A diesel, water-cooled, 6-cylinder, 4-cycle engine. The larger sizes produce up to 500 kW. (Onan)

conversion in power is made (transformers step down voltage). The plant may then be kept in reserve for emergency standby, or other service. Such versatility makes the AC power plant unique.

An engine-driven AC plant with a DC starting system can supply a limited amount of DC power as well. It would be of 12-volt or 32-volt design, depending on the starting characteristics of the plant.

A plant must be operated whenever power is needed. There is an interruption in power when the plant needs repair or service. Such a problem is usually resolved through proper care and maintenance.

Mechanical Energy Sources

The previous discussion of electrical power plants covered their electrical-generator component. Now let us review the various types of generator driving power that are often used. The prime source of mechanical power for driving the generator is usually one of these three types: gasoline, diesel, or gas engine. Fig. 2-5.

A high degree of efficiency has been reached in the design and production of these internal combustion engines. This insures that any one of the three types will produce a dependable source of power.

All engines must maintain a specific generator speed to develop the desired frequency. For example, 60 hertz (Hz) requires 3600 RPM, or some fraction ($\frac{1}{2}$ or $\frac{1}{3}$) of 3600, which would be 1800 or 1200. For 50 hertz, this must be 3000 RPM or 1500 or 1000.

• *Gasoline engine.* The gasoline engine is probably best known, since its basic design is the same as that of the engine used in automobiles. Sometimes, however, there may be a low requirement for power in some plants. Here, a gasoline engine may need only one or two cylinders, instead of four or more, as required in an automobile.

Most of the engines in power plants are of the four-cycle type that operate either at 1800 or 3600 RPM. Some are two-cycle engines, operating at 3600 RPM. These are

2-5. *Gas and gasoline-type engine's intake on the left; diesel engine's intake on the right.* *(Onan)*

used for extremely light-weight, portable units. Because these engines present certain service problems, the trend in design is toward four-cycle engines which run at moderately high speeds. By doing this, more generating capacity is obtained, with less plant weight. Older engines ran at 300 or 450 RPM. Modern engine design has increased the speeds. Now, 1800 RPM is established as standard, and 3600 RPM is universally accepted. Fig. 2-6.

All plant engines have compression ratios designed to take full advantage of high-octane gasoline. This gives greater operating efficiency.

• *Diesel engine.* The diesel engine is also of the internal combustion type. It uses the heat of compressed air to ignite the fuel, instead of an electric spark as the gasoline and gas engines do. Production problems, very high initial costs, and limited use prevented early customer acceptance of the diesel. Today, diesels are in demand. An

2-6. *This gasoline engine produces 13 HP at 1800 RPM. It has remote start-stop capability with 12 V, DC. The generator puts out 120 V, 1 ∅, 60 Hz, at 4 kW. Water-cooled model shown here. (Onan)*

2-7. *A diesel engine-single cylinder, 4-cycle, 5.7 HP, air cooled. The generator puts out 3 kW of 120/240 V, 60 Hz, 1 ∅, AC. This engine can be used for emergency home supply. (Onan)*

important reason for their widespread use is their ability to run on comparatively low-cost fuel, their low operating costs, and their long life.

Two combustion cycles are available in diesel engines: two-cycle and four-cycle. Both types are practical. However, four-cycle engines are sometimes more economical. They use less fuel than two-cycle engines.

Due to extremely high combustion pressures, all components used in this engine must be larger than those of gasoline engines. They must be made of very strong material. For example, the injection pump times and measures the amount of fuel for each cylinder. This part must have very close tolerances for good starting and efficient operation. This adds to the expense of the engine. Fig. 2-7.

● *Gas engine.* Natural and manufactured gas (or a combination of both) and bottled, or LP (liquified petroleum), gas are widely used as engine fuels. A standard gasoline engine, equipped with special carburetion, permits the use of such fuel. In theory, combustion is the same, except the fuel is already in a gaseous state. Thus, the process of gasoline vaporization is eliminated. Use of gas fuels allows the engine to run cleaner, with lower maintenance costs as a result.

Natural gas has a high enough Btu content (British thermal unit; SL metric system: joule, J) to permit the engine to develop near-rated horsepower. Manufactured gas, or a mixture of it and natural gas, both have a reduced Btu content. This lowers the output of the engine and plant. The plant may require *derating* if the fuel used has a Btu content of 1000 or less. Derating means lowering the plant output because the fuel does not contain enough energy to meet the standard fuel value.

Bottled, or LP, gas is usually *butane* or *propane* or a mixture of both. It is supplied according to local and climatic conditions. It is stored in tanks in liquid form under pressure. When pressure is reduced, the liquid becomes a gas and operates the same as natural gas. In all cases, the Btu content is high. The engine output is comparable to that obtained from gasoline.

Economical Considerations of Various Engines

Because of the apparent lower operating costs of diesel engines using oil for fuel, the person responsible for ordering plant machinery might be inclined to specify that type of engine. There are, however, several factors to consider.

• *Original cost.* Diesel plants cost more than either gasoline or gas engines for several reasons. There is not the high mass production, which would help reduce manufacturing costs. More expensive components are required for diesel engines. Because they are heavier than other engines, transportation costs are higher for them. Gas plants are slightly higher in cost than gasoline plants.

• *Installation cost.* For an average job, installation costs are about the same for the three types of plants. But when a gasoline plant must meet local fire codes for a building installation, the cost of gasoline plants may be the highest of the three plant types. Strict regulatory codes exist because volatile fuel is being stored within the building and adequate protection must be provided. Installation of gas engine plants is probably the least expensive, with the diesel rating next in economy. The same comparison would probably apply to insurance on a building. Check insurance policies to make sure, however.

• *Starting methods.* Gasoline and gas engine plants are easier to start and are manufactured with more types of starting systems than diesel plants. Fig. 2-8. Diesels can be hand cranked, generally in sizes up to 3 kilowatt. Other types can be hand-cranked in all sizes. All plants can be electrically started. Smaller gasoline and gas plants may be started through the exciter winding of the generator. Larger sizes may be started by an automotive-type starter. Only small diesel plants can be exciter cranked. All diesels can be electrically cranked, however. The ambient temperature (air temperature surrounding the plant) must be above zero Fahrenheit to assure starting. This may be accomplished with glow plugs, air heaters, or room and oil base heaters.

There are many types of starting methods and variations of those methods. Let us consider the five most common methods:

1. *Manual.* This plant, using magneto (permanent magnets) ignition, is manually started by hand crank or rope. It is stopped by a push button. This method is used to start portable, mobile, intermittent, and sometimes, emergency-service units. Fig. 2-8.

2. *Electric.* This plant, using magneto or battery ignition, is electrically started or stopped by means of push buttons on the unit. Stopping is possible from any push

2-8. *Starting methods for power plants. Left to right: hand crank, switched remote-control, and automatic load transfer. (Onan)*

button located up to 250 feet [80 metres] from the plant. Storage batteries furnish current for cranking. Fig. 2-8.

3. *Remote control.* This type of plant is electrically started or stopped by push buttons mounted on the plant, or from any number of push-button stations located up to 250 feet [80 metres] from the plant. Fig. 2-8. A set of automotive-type batteries supplies the electric power for cranking. These batteries are kept charged by a special circuit, controlled at the plant. Depending on the voltage and capacity of the batteries, the starting circuit may also be used to supply a small amount of DC current for lighting or other purposes.

The remote control is practical for many installations. When used with many start-stop stations, it provides almost the same service as an automatic plant. However, its operations are controlled. This eliminates haphazard operation and results in greater efficiency. A water pump cooling system may also be run with this method. However, automatic heating or refrigeration requires a full-automatic starting method. To convert from remote control, all that is necessary for most AC remote control plants is the addition of a full-automatic control panel designed for this purpose.

4. *Full automatic.* This plant starts automatically whenever an electric-lighting load of about 75 watts (resistance load much higher) is turned on. It continues to run until the load is turned off. This starting method will operate a refrigerator, or any properly connected automatic equipment, without any individual attention. Basically, this method uses a standard remote control plant, plus an automatic control that makes the plant fully automatic in starting. Thus, a remote-control plant may be converted at any time to full-automatic starting by the addition of the panel.

A plant with this automatic-starting method is likely to run more frequently than a plant with other types of starting. Any minimum electrical load will start it.

Special consideration must be given to the size and use of a full-automatic plant. If several automatic devices are to be used, with a possibility that they may all operate at one time, the plant must be large enough to handle the total starting and running load. A plant using the other starting methods may be smaller in size because its load and usage can be controlled.

5. *Automatic load transfer.* This starting method is designed for emergency standby plants. Fig. 2-8. Whenever there is a power interruption on a commercial line, an automatic transfer disconnects the load from the line. It starts a standby engine and keeps the plant running. It connects the load to the plant, then reverses these procedures when normal power is resumed. It ordinarily supplies a trickle charge to the starting battery when the plant is not running.

An automatic load transfer may be installed and connected to any AC remote-control plant to provide this type of starting. The plant and transfer mechanism may be connected in several ways. One method is to allow the combination of the two to handle the entire load. Another solution provides that the plant and transfer handle only a certain portion of the load on a preselected circuit. The size of both plant and transfer is selected according to electrical requirements.

Certain relays may be added to the transfer to service special needs as required. These include voltage-sensitive relays and time-delay relays. Voltage-sensitive relays will start the plant when line voltage varies above or below a predetermined value. Time-delay relays delay starting and stopping. The plant does not start on momentary power outages nor shut down before high-line power is securely restored. Where the

2-9. *Air-cooled (left) and water-cooled (right) engines. (Onan)*

starting load is larger than plant capacity, time-delay relays are necessary to transfer the load to the plant in stages so that motors can be started one at a time.

● *Operation.* The main point to consider is the cost and availability of the fuels. In almost all cases, diesel fuel costs are the lowest, with gasoline and gas being about the same for similar Btu value. Prices, of course, are subject to change. Exceptions in fuel economy occur where a plant has access to a free gas well. This would obviously give it the lowest fuel cost when compared to other types of plants. However, if this plant operates only a few hours per year, the savings on fuel costs would not offset higher initial equipment costs.

Availability of fuels is of prime importance. A standard gasoline at somewhat stable prices is available almost everywhere. However, the other fuels, gas and diesel, cannot always be secured. This should be a determining factor in selection of the type of engine for a plant.

Lubricating oil used in the engine crankcase may be slightly more expensive per quart for the diesel plant. The small amount required, as compared to the operating fuel, does not make it an important price factor.

Gasoline and gas plants require about equal expense for engine maintenance. Such costs may be lower than for the diesel engine, even though the latter has fewer points to service. An operator of a gasoline, gas, or diesel plant can, in most cases, take care of his or her own maintenance. However, some diesel engines may require special training for servicing and maintenance personnel. This type of specialized service has become more readily available. It does add to the overall cost of diesel plant operation.

● *Overhaul.* All engines require a periodic overhaul, with carbon removal and valve grind. In addition, certain parts require replacement, depending on their type and length of service. A complete overhaul for gasoline and gas plants will probably cost less than for diesel plants. (Such an overhaul, however, will be required more often than for the diesel plants.) As a result, total overhaul costs for the life of a diesel plant may be lower than for other types because most service to a diesel plant is usually for accessory equipment.

● *Engine cooling.* Two basic methods are used for the job of cooling an engine that develops a large amount of heat in its operation: *air* and *water.* Fig. 2-9.

Water-cooled engines have radiators that are supplied by municipal water. This usually raises costs higher than for air-cooled systems. The single-duct radiator heat exhausting system has been equaled by new air-cooled systems.

It may be necessary someday for you to operate, or at least identify, these power plants. The information presented should be sufficient for you to detect advantages and disadvantages. This information might enable you to recommend a particular unit.

FALLING-WATER GENERATORS

The principle of electricity generation is the same, whether it uses mechanical energy produced by gasoline or diesel engines, or falling-water or nuclear energy. In this part of the chapter, you will learn about the production of electricity by falling water. This method uses the energy released by the force of water falling over a distance to turn a generator and produce electricity.

2-10A. *Hydroelectric generating plants at Niagara Falls, New York. (New York State Power Authority)*

Niagara Power Project

The Niagara Power Project is an example of falling water used to generate huge amounts of electrical energy. As the description of the operation and utilization of the plant progresses, you will probably realize the similarity of this type of power generation to that of fossil-fuel and nuclear power plants. Although these different types of generators of electricity are the same, the method for generating mechanical energy is different. (Mechanical energy is needed to turn the shaft upon which a coil of wire is revolved in a magnetic field.) Initial cost of building a power plant of this magnitude is extremely high. However, the system is comparatively maintenance free for about one hundred years.

Niagara is one of the largest hydroelectric developments in the western world. Total installed power at Niagara's two generating plants (Robert Moses and Lewiston) is 2 190 000 kilowatts. Fig. 2-10. There are 340 miles [544 km] of transmission lines to interconnect with the Niagara Project and the dam at St. Lawrence.

Switchyard

The Niagara switchyard is situated on a 35-acre [14-hectare] site south of the power canal, halfway between the Robert Moses and the Lewiston power plants. Purpose of the switchyard is to collect and meter power from the generators and send it out

2-10B. *Robert Moses power plant in Niagara Falls. (New York State Power Authority)*

over transmission lines. The switchyard has three voltage sections: 115 kV, 230 kV, and 345 kV.

Power enters the switchyard from the Robert Moses plant through seven 115-kV cable circuits, and six 230-kV cable circuits. It enters from the Lewiston plant by means of four 230-kV circuits. The cables are installed in underground power tunnels.

NUCLEAR POWER PLANTS

A nuclear power plant is in many aspects similar to a conventional fossil-fuel burning plant. The chief difference is in the way the heat is generated, controlled, and used to produce steam to turn the turbine generator. Fig. 2-11.

In a nuclear power plant, the furnace for burning coal, oil, or gas is replaced by a reactor that contains a core of nuclear fuel. Energy is produced in the reactor by a process called *fission*. This fission process splits the center, or nucleus, of certain atoms when they are struck by a subatomic particle called a neutron. The resulting fragments, or fission products, then fly apart at great speed. They generate heat as they collide with surrounding matter.

The splitting of an atomic nucleus into parts is accompanied by the emission of high-energy electromagnetic radiation and the release of additional neutrons. The released neutrons may in turn strike other fissionable nuclei in the fuel, causing further fissions.

A nuclear reactor is thus a device for starting and controlling a self-sustaining fission reaction. The nuclear core of the reactor generally consists of fuel elements in a chemical form of uranium and thorium, or plutonium, depending on the type of reactor. Heat energy is produced by the fissioning of the nuclear fuel. A coolant is then used to remove this heat energy from the reactor core so that it can be used in producing electricity.

Water-Cooled Reactor

In a water-cooled reactor, the nuclear fuel, which has been compacted into uniform pellets, is placed into tubes or fuel elements. These fuel elements are then sealed at the top and bottom. They are arranged by spacer devices into bundles called fuel assemblies. The spacer devices separate the fuel elements to permit coolant to flow around all of the elements. This process is necessary to remove heat produced by the fissioning uranium atoms. Scores of fuel assemblies, precisely arranged, are required to make up the core of a reactor. Fig. 2-12.

This geometric arrangement is necessary for several reasons. Nuclear fuel, unlike fossil fuel, has a very high energy density — that is, tremendous quantities of heat are produced by a small amount of fuel. Therefore, the fuel must be arranged as stated above to permit the coolant to carry away the heat. To accomplish this, the fuel must be dispersed, rather than lumped together in a large mass.

Chemical reactions between the fuel and the coolant must be avoided. Thus, as a safety precaution, radioactive materials produced must be enclosed. For these reasons, the fuel is contained in individual tubes, or fuel elements. The cladding material, from which the tubes are made, must meet rigid specifications. It must have good heat-transfer characteristics. It must not react chemically with either the fuel or the coolant. Also, it must not absorb the neutrons produced in the fissioning of the fuel at a rate that could be detrimental to the chain reaction. The cladding material generally

2-11. A conventional fossil-fuel power plant. (Edison)

2-12. A boiling-water reactor. (Edison)

55

used for this purpose is thin-walled stainless steel. An alloy of the element zirconium can also be used.

Gas-Cooled Reactor

A gas-cooled reactor, which utilizes the inert gas helium as the coolant, has a different core structure than the water-cooled reactor. The fuel elements are made of graphite. Graphite acts as the structural material and the neutron moderator. It also serves as the cladding material. Nuclear fuel consists of both uranium and thorium. It is formed into the center of the fuel element. Because the helium is inert, the graphite serves as cladding for the nuclear fuels. (An inert gas will not react with nor corrode the graphite or any other structural material.)

Physically, the fuel elements for a gas-cooled reactor are much larger than those of a water-cooled reactor. They are not bundled into individually arranged fuel assemblies and spaced for the circulation of coolant as are those in water-cooled systems. Despite their larger size, however, several hundred fuel elements are required to make up the core of a reactor. Fig. 2-13.

Nuclear fuel, whether in the form of fuel assemblies for a water-cooled reactor or fuel elements for a gas-cooled reactor, is placed in the reactor to produce heat. The heat in turn is converted into electricity. After several years of operation, a nuclear core must be replaced. With the passage of time, the absorption of neutrons by the accumulated fission products is so great that there are too few neutrons remaining to maintain a chain reaction. Fig. 2-14.

After fuel assemblies, or elements, are removed from the reactor, they retain some material. The material must be reclaimed because of its economic value. In fact, only about 1 to 3% of the uranium has been "burned" in nuclear reaction. The remaining 97 to 99% is locked in the hundreds of fuel elements of the "spent" core. Figs. 2-15 and 2-16.

The procedure by which the nuclear fuel is reclaimed is called *reprocessing*. The first industry-owned reprocessing plant began operation in 1966.

FOSSIL-FUEL POWER GENERATORS

The steam needed for driving turbines (which in turn drive electric generators) must be produced by heat. The method of heat production often becomes a rather difficult engineering problem. With the development of some dependable sources of heat from fossil fuels, former design problems have been simplified. Almost any substance may be used as a fuel. If it can be pulverized and fed into furnaces with extremely high temperatures, it will burn.

Coal, Natural Gas, and Oil

Coal has been one of the most abundant sources of energy. It is used frequently to produce electricity by providing steam in those power plants designed for the use of coal. It is not, however, the only fossil fuel used to produce steam for power plants. Natural gas and oil are also used. There are some problems associated with these fuels. The waste products can pollute the atmosphere. Combustion products discharged into the atmosphere from a steam generator must meet EPA (Environmental Protection Agency) emission standards for particulate, sulfur oxides, and nitrogen oxides. Clean fuels for a steam generator may be defined as those fuels whose com-

2-13. A high-temperature, gas-cooled reactor. (Edison)

STEAM GENERATOR

HELIUM CIRCULATOR

TURBINE-DRIVEN CIRCULATOR

HELIUM, 700-PSI

REACTOR CORE

1430°F

760°F

TURBINE GENERATOR

CONDENSER

FEEDWATER HEATERS

PUMP

TRANSFORMER

2-14. A pressurized-water reactor. (Edison)

REACTOR

CONTROL RODS

PRESSURIZER

STEAM GENERATOR

PRIMARY CONTROLLED LEAKAGE PUMP

TURBINE GENERATOR

FEEDWATER HEATER

CONDENSATE PUMP

CONDENSER

TRANSFORMER

2-15. *A gas-cooled, fast-breeder reactor. (Edison)*

2-16. *A liquid-metal, fast-breed-er reactor. (Edison)*

bustion products do not require stack-gas cleanup to meet the limits set by antipollution agencies.

SYNTHETIC GAS. Steam generation from synthetic low or medium Btu gas is a development probably several years in the future. This is due to the lack of availability of a commercial gasifier system at this time. Conversions of existing units to gas must be handled on an individual basis, to ensure operation within original design limitations.

COAL-FIRED POWER PLANTS. Figure 2-17 shows a Niagara Mohawk power plant in the town of Tonawanda, New York, located on the Niagara River. A huge pile of coal is

2-17. *Niagara Mohawk's coal-fired plant located on the Niagara River supplies power for Buffalo, New York. (NMP)*

needed to fire the furnaces. The river here also provides an ample supply of water. Figure 2-18 shows the inside of the power plant. Here, the turbines, fed by steam, rotate electrical generators to produce the power needed.

Figure 2-19 shows an operator in the control center of a large nuclear power plant. The control centers of coal-powered plants are somewhat similar in appearance. In both, operation is by remote control. Monitoring charts are checked constantly for performance and demand.

59

2-18. *Generator room where steam turbines drive the alternators to produce electricity. (NMP)*

2-19. *The power control center in a large nuclear power plant. (Courtesy U.S. Department of Labor)*

QUESTIONS

1. What is meant by battery polarity?
2. What is an armature?
3. What is another name for an AC generator?
4. At what speed must a gas-driven generator be turning to produce 60 Hz AC?
5. Which engine is the most economical to operate in terms of using it to generate electricity?
6. What are three starting methods for engine-driven generators?
7. What are some of the safety problems associated with nuclear plants that are used for power generation?
8. How is a gas-cooled nuclear power plant different from one that uses a water-cooled reactor?
9. What is another name for a fossil-fuel generator?

CHAPTER 3

Distributing Electricity

URBAN AND SUBURBAN DISTRIBUTION

Electricity may be generated by fossil-fuel power generators, by atomic-powered generators, or by the use of falling water to drive generators. It is also possible to generate power by engine-driven generators. In Alaska, for example, long distribution lines are impractical due to ice conditions that would result in line damage during storms. In Alaska, engine-driven generators are common.

Once electricity is generated in sufficient amounts for consumption in large quantities, the second necessary step is to get the energy to the consumer. Herein lies a distribution problem: that of stringing and maintaining long lines.

This chapter will be concerned with some of the problems and methods of distributing electricity after it is generated by public utilities.

Figure 3-1 diagrams the process of getting power from the generating plant along high-voltage transmission lines, to voltage step-down substations, and then to office buildings, stores, apartments, and large factories. Further reductions are necessary in voltage to reduce the power to proper voltages (120-240) for home use.

A main concern of most utility companies is to get the greatest possible amount of usable energy to their customers. Voltage losses on the way from the plant to consumers are minimized by proper use of step-up and step-down transformers, and the correct size of wire.

Electrical Power

Most electrical power is generated as three-phase (3∅). It is stepped up to 132 000 volts, 238 000 volts, or even to 750 000 volts. The frequency is 60 hertz. Sometimes, however, 25 hertz AC is generated for use by some consumers who still may have older equipment that uses this frequency. Motors, and other equipment using this lower frequency, would be too expensive to replace. However, most 25-hertz equipment, when worn out, is replaced with 60-hertz equipment. A separate distribution system is necessary to send 25-hertz power to its destination. This is expensive.

UNITED STATES POWER COMPANIES. In the United States, power is generated by shareholder-owned electric companies, by government-owned facilities such as TVA (Tennessee Valley Authority), and by privately owned companies. The most popular system in the United States is the shareholder-owned. In this system, an electric company is owned by people who buy shares in the company on the stock exchange.

Shareholder-owned electric companies differ in two major respects from commercial and manufacturing companies:

 • Their activities are confined to geographic service areas established by regulatory agencies.

HOW ELECTRICITY IS MADE AND BROUGHT TO THE USERS.....

3-1. The generation and distribution of electricity.

● They operate free of competition from other electric companies as long as the best interests of the public are served.

Competition by suppliers of other forms of energy, such as gas, coal, and oil is intensive, however.

Electric utilities are granted exclusive franchise areas because of the extremely large investment required of them to supply electricity, compared to the investment necessary for other types of businesses.

CANADIAN POWER COMPANIES. In Canada the power companies are owned and controlled by the provinces. Because equipment standards are set by the provinces, companies can be very particular concerning the equipment used on their lines. In addition they can specify and regulate inspection of devices hooked to the power lines (such as home appliances). In this way the power companies, along with the Canadian Standards Association, are able to demand and get safer manufactured devices in homes and industry. Their regulations can specify whether equipment is safe to operate. The Canadian companies can also designate the type of wiring for a house, business, or industry.

HIGH-VOLTAGE TRANSMISSION. Figures 3-2 through 3-5 illustrate some of the devices and equipment used to get the high voltage from the generator to the customer.

3-2. The distribution network for a large generating plant. (NMP)

3-3. The control room of a large generating plant. (NMP)

63

3-4. Installing a new 238 000-volt transmission line. (NMP)

3-5. Note the size of the transformers in comparison to the man nearby. This is a high-voltage substation. (NMP)

Local Distribution

Figures 3-6 through 3-15 are concerned with transferring power from the substation to the local user. Most of these photos show overhead installations of power distribution systems. There are also a number of undergound installations. Large cities and most suburban areas are now demanding removal of unsightly wires and poles. Underground wiring, although more expensive, is often used.

UNDERGROUND SYSTEMS. Underground systems for distributing electrical energy are in the middle of a revolution that began about ten years ago. A prime objective has been to reduce the cost of these systems so that they could be used in areas previously served only by overhead systems. Aluminum, long known as an economical electrical conductor, has contributed to this revolution.

3-6. *A cross-sectional view of a high-voltage underground cable. Note the various materials used for insulation. (NMP)*

3-7. *Stringing new cables for local service. (CILCO)*

3-8. *Providing telephone service to an apartment complex. In most neighborhoods, telephones and electrical power use the same poles. (Courtesy U.S. Department of Labor)*

3-9. *Laying wires in a trench for an underground distribution system. These wires will be used to provide underground power circuits for homes in a subdivision. (NMP)*

3-10. *A buried transformer being spliced into service. Notice how it is buried in an enclosure. (NMP).*

65

3-11. Above-ground distribution system for homes. The transformer is connected to the top wire for high-voltage input. The three wires below the transformer are the low-voltage (240-volt) output that goes to the house.

3-12. The other end of the wires from the transformer are brought to the house and anchored here.

3-13. Wires from the pole are spliced to a larger three-wire Romex cable to carry power to the meter and then down to the distribution panel in the basement.

3-14. Close-up of a box containing a meter. The kilowatt-hour meter can be removed from the plug-in socket if necessary.

3-15. *This is a five-in-line clock dial, single-phase, 240-volt, three-wire, kilowatt-hour meter. (Sangamo)*

Actually, underground distribution is as old as the electric power industry itself. In 1882, Thomas A. Edison built the first central power station on Pearl Street in New York City. He installed feeders and mains under the city streets to distribute the power generated in the station to its customers. These were pipe-type cables, with a pair of copper conductors which were separated from each other, and from the enclosing pipe, by specially designed washers. The pipe was then filled with an insulating compound. This compound consisted of asphaltum boiled in oxidized linseed oil, with paraffin and a little beeswax added.

The development of lead-sheathed insulated cables that could be pulled into previously installed ducts greatly simplified construction of underground distribution systems. However, the high costs of such systems remained. As a result, for many years underground distribution was limited largely to the central portions of cities. Here load densities are high and the congestion that would result from overhead systems is highly undesirable. Fig. 3-6.

Today, costs of underground distribution have been reduced to levels that are economically practical for service to light- and medium-density load areas. Underground systems now are being installed in most new residential areas. Home buyers in these areas generally agree that the improved appearance and enhanced value of their property more than justify the added cost of underground service. Figs. 3-9 and 3-10.

Some utilities are replacing existing overhead systems with underground systems in residential areas, particularly where load growth requires added capacity. Even some rural systems are being installed underground today.

Reduced costs of underground systems have been achieved through the development of efficient new materials and low-cost system components. Progress has also been made in simplifying construction techniques and installation methods. One of the most important developments has been new synthetic insulating material, notably thermosetting, or cross-linked polyethylene. Cables with this type of insulation may be buried directly in the earth without expensive extra duct work. The development of improved cable plowing and trenching equipment has also helped to reduce installation costs.

RURAL ELECTRICITY

The industrial revolution of the nineteenth century, which had transformed living styles in cities the world over, scarcely touched life on the farms of the United States. As a result, the American farmer, at the dawn of the present century, was earning his living in a way that had changed little from that of the first colonists who settled along the Atlantic seaboard. The tools he used were simple and ancient: the wheel, the lever, the block and tackle, and the plow. For most tasks, the farmer could draw only on his own strength, or that of horses and other animals. The children studied by the dim light of a kerosene lamp. The women were slaves to the wood stove and washboard.

Federal Agencies

In the United States, the federal government does not provide retail electric service over distribution systems, either to urban or to rural areas. The role of government in the electric power industry is fourfold:

1. The Federal Power Commission exercises licensing authority over the utilization of hydroelectric sites on navigable rivers of the country. It also maintains certain controls over the interstate transmission and sale at wholesale of electric energy generated by electric companies.

2. The Bureau of Reclamation, the Army Corps of Engineers, and the Tennessee Valley Authority build and operate some of the nation's hydroelectric generating plants. The TVA also builds steam plants.

3. The Bureau of Reclamation, TVA, the Bonneville Power Administration, and similar agencies build and operate transmission lines for marketing power wholesale.

4. The Rural Electrification Administration (REA) makes loans for rural electrification, including generation, transmission, and distribution systems.

Electricity for Everyone

The notion that electricity generated at a central station could be distributed to every farm in the United States took hold of people's minds slowly. Electric service was theoretically within the reach of rural families when the discovery was made that alternating-current voltage can be "stepped up" for transmission and "stepped down" for utilization. This of course, means that power can be delivered economically to areas that are distant from the generating station.

Of course, technological theory was not sufficient to bring electrical power to rural areas. Financing on a large scale had to be provided. In the United States, farmers live on the land they cultivate. Farmhouses, therefore, are widely scattered across the countryside. In some ranch areas of the western states, houses are many miles apart. In such thinly populated rural areas, electric companies could see little prospect for profit. Farmers usually were required to pay construction costs of individual line extensions to provide service. Rates were high. Most of them could not afford this and, therefore, remained without electricity.

REA

In 1935, President Franklin D. Roosevelt created the Rural Electrification Administration as an emergency relief program. In May, 1936, the Congress of the United States passed the Rural Electrification Act. This established REA as a lending agency of the federal government. It had the responsibility of developing a program for rural electrification. The act authorized and empowered REA to make self-liquidating loans to companies, cooperatives, municipalities, and public power districts. These loans were to finance the construction and operation of generating plants, transmission and distribution lines, and related facilities for the purpose of furnishing electricity to persons in rural areas. Fig. 3-16.

In addition to making loans, REA furnishes technical assistance to its borrowers. It offers advice on engineering, management, accounting, public relations, power use, and legal matters. REA, however, does not construct, own, or operate electric facilities.

RURAL DISTRIBUTION OF POWER. A typical rural distribution system serves a relatively large geographical area, usually with a uniform, light-load demand. An average system serves 5000 consumers with 1500 miles [2400 kilometres] of distribution line. However, some rural systems serve over 25 000 customers. Here, farms are quite large, generally averaging 300 acres [121 hectares] or more. Typically, about three farms are served by each mile of distribution line. Loads are predominately single-phase, although a few three-phase installations may be located within the area. Towns, villages, and cities are served by separate distribution facilities not financed by REA. The typical rural system load is growing rapidly. It requires good voltage regulation and a high level of service continuity. Fig. 3-17.

These factors have a major influence on design and construction of a rural distribution system. The preponderance of single-phase loads has led to systems that are essentially single-phase in construction and operation. Approximately 80% of the lines

3-16. Dairy farmers use electric cow milkers. (Courtesy U.S. Department of Labor)

3-17. Today's farmer often has use of a complete workshop due to the availability of inexpensive electrical power. (REA)

being used are single-phase. The system that has evolved for rural service is typically an overhead multi-grounded, neutral, radial system on wood poles. It uses relatively small conductors that are usually aluminum with steel cores (ACSR).

The most common voltage is 7200 volts phase-to-ground, or 12 470 volts phase-to-phase. There are some sparsely settled areas remote from power sources where the individual farm loads are heavy. For such areas, it has been more economical to use 14.4/24.9 kilovolts as a distribution voltage.

Compared to other types of systems, the multi-grounded neutral system offers a number of distinct advantages for serving the remote area just described. Long-span, single-phase branches may be constructed most economically in vertical configuration. This is the basic line-supporting structure used in United States rural electrification.

- *Telephone and power lines.* The experience of rural distribution systems operating in close proximity to telephone or other communication systems has been satisfactory. The two systems frequently parallel each other at roadside separation. In addition, a large amount of line is built with the telephone and electric systems sharing the same poles. No special construction is required for the electric distribution system. However, the transformers used are built to a specification that controls the harmonic content of the excitation current. Telephone systems in the United States have determined that the most satisfactory method of operating is to adopt construction standards that are characterized by a very low susceptibility to power line influence. These standards make it unnecessary for telephone companies to request special construction from power systems to reduce power line influence on their communication cables.

SINGLE-PHASE POWER. The principal reason for favoring single-phase rural distribution lines is that most rural loads are single-phase. An average farm has no three-phase requirements. Most farm motors, including those used for feed grinding, are below 5 horsepower. Larger and deeper irrigation wells, however, frequently require pump motors of 25 horsepower or more, for which three-phase service is provided. In the past, 7½ horsepower generally was regarded as the largest practical single-phase motor. The cost of this motor was about double that of a comparable three-phase motor. Fig. 3-18.

Several hundred dollars for an occasional motor seems preferable to thousands of dollars for three-phase lines. Meanwhile, manufacturers are becoming interested in larger single-phase motors that will perform as well as, yet cost no more than, three-phase motors. It also should be noted that many rural systems have the capacity to serve 30-horsepower, single-phase motors without undue flicker problems. In fact, single-phase loads no longer are limited to any specific size. The present practice is to allow the capability of the system, at the time of installation, to determine the maximum size of the single-phase load that can be served.

INSTALLATION OF SERVICE BY POWER COMPANY

You have, no doubt, noticed that there are some standards for the transmission of electrical power over distances. The towers, poles, and supports of various types are standardized as to size and treatment to endure years of exposure. There are some very definite minimums and maximums for the crossing of roads and various other

DIVISION OF OWNERSHIP

CUSTOMER — COMPANY

CUSTOMER
METERING CABLE —

1 1/2'' WEATHERHEAD

LEAVE CUSTOMERS CONDUCTORS 24" LONG
COMPANY WILL MAKE ALL CONNECTIONS TO
SERVICE DROP. BOND STEEL CONDUIT TO
NEUTRAL

SERVICE BRACKET FURNISHED BY
COMPANY - INSTALLED BY CUSTOMER

METERED LINES TO
CUSTOMER'S LOADS.
SERVICE EQUIPMENT
AT EACH BUILDING.

COMPANY'S SERVICE DROP

TWO CURRENT TRANSFORMERS MOUNTED
ON BRACKET FURNISHED BY COMPANY
AND INSTALLED BY CUSTOMER.

CUSTOMER'S METER POLE

FURNISHED AND INSTALLED BY CUSTOMER

CONSULT COMPANY FOR POLE AND
GUY REQUIREMENTS

POLE TO BE 5"MINIMUM DIAMETER AT TOP,
8"MINIMUM DIAMETER 6 FEET FROM BUTT.
NORMALLY 35 FOOT POLE EXCEPT AS
OTHERWISE NEEDED FOR SERVICE DROP
HEIGHT REQUIRED

1 1/2"GALVANIZED RIGID STEEL OR ALUMINUM
CONDUIT FURNISHED AND INSTALLED BY
CUSTOMER

POLE TO BE TREATED WITH PENTA-
CHLOROPHENOL IN OIL-EEI SPEC. TD 100,
OR ACCEPTABLE EQUIVALENT.

STRAPS AT NOT MORE
THAN 4 FT. INTERVALS.

METERING CABLE FURNISHED
BY COMPANY.

WHERE SPAN FOR SERVICE DROP
EXCEEDS 30 FOOT THE POLE MUST
BE GUYED FOR 1000 LBS. STRAIN.

WEATHERTIGHT
FITTING

METER ENCLOSURE FURNISHED BY
COMPANY AND INSTALLED BY CUSTOMER

GROUNDING CONDUCTOR
#4 COPPER MINIMUM WITH
WOOD SHEATHING FOR ITS
FULL LENGTH ON POLE

4'
to
6 1/2'

5/16" GUY STRAND

20' MINIMUM

6'-0" MINIMUM

5/8" x 8'
GROUND
ROD

3'
MINIMUM

130 SQ. INCH
ANCHOR

3-18. *A farm meter pole. Single-phase, three-wire, 120/240 volts for loads exceeding 40-kilowatt demand. Some farmers purchase power from utility companies other than REA. (NMP)*

8'-0" MINIMUM OVER FLAT ROOF
3'-0" MINIMUM IF ROOF CAN NOT BE READILY WALKED UPON

10'-0" MINIMUM

18'-0" MINIMUM OVER PUBLIC STREET

12'-0" MINIMUM OVER RESIDENTIAL DRIVEWAYS
18'-0" MINIMUM OVER INDUSTRIAL AND COMMERCIAL DRIVEWAYS

3-19. *Minimum vertical clearance of service drops below 600 volts. (NMP)*

obstacles on the way from the generator to the user. Figure 3-19 shows the minimum clearance of service drops below 600 volts.

Figure 3-20 shows the attention given to the service entrance riser support on a low building. This work is usually done by the electrician, and at least 24" service conductors are left for connection by the power company when it brings power up to the house or building.

Most new suburbs are demanding that electrical service be brought in from the line to the house by underground cables. Figure 3-21 shows the requirements for such an installation. Note how far the meter is mounted above ground. This service shows the pole in the rear of the house. Some localities also require that the entire electrical service to an area be underground. This eliminates poles in the rear of the house.

Meter Installations

The electrician is usually required to place the meter socket trough. It is usually supplied by the power company and installed by the customer. Figure 3-22 shows a typical meter installation with a socket for plugging in the meter once the service has been connected. In the case of apartment houses it will be necessary in some cases to install two or more meters.

Note in Fig. 3-23 (Page 76) that six more meters can be installed indoors in a single-phase, three-wire, 120/240-volt service or single-phase, three-wire, 120/208-volt ser-

WEATHERHEAD

REDUCER

LEAVE SERVICE CONDUCTORS 24" LONG FOR CONNECTION TO SERVICE DROP BY COMPANY

SERVICE BRACKET AND MOUNTING STRAP FURNISHED BY COMPANY INSTALLED BY CUSTOMER

9"

COMPANY'S TRIPLEX SERVICE DROP 1000 LBS. STRAIN

BACKBRACE

2' - 0" MAX. WITHOUT BACK BRACE

THE RISER SHALL BE CAPABLE OF WITHSTANDING A HORIZONTAL PULL OF 1000 LBS. AT THE SERVICE DROP ATTACHMENT. PROVIDE BACK BRACE WHERE NECESSARY.

BOND RISER PIPE TO SERVICE NEUTRAL

APPROVED VENT PIPE FLANGE

3/8" U BOLT AS CLOSE TO ROOF AS POSSIBLE

ALTERNATE PIPE OR CONDUIT LOCATION (IN EXTERIOR BUILDING WALL ONLY.)

2" OR 2-1/2" GALVANIZED STEEL PIPE OR RIGID CONDUIT

WHERE PIPE IS USED, SERVICE ENTRANCE CABLE IS REQUIRED WHERE CONDUIT IS USED SEPARATE CONDUCTORS MAY BE USED.

ALTERNATE SERVICE EQUIPMENT

3/8" U BOLT RISER SUPPORT

ALTERNATE SERVICE ENTRANCE

WEATHERTIGHT FITTING

METER SOCKET TROUGH FURNISHED BY COMPANY INSTALLED BY CUSTOMER

INSULATING BUSHING

PROVIDE VERTICAL SURFACE FOR MOUNTING METER SOCKET TROUGH

MAXIMUM LENGTH OF UNGUARDED SERVICE ENTRANCE CONDUCTORS WITHIN WALL SHALL BE 12"

4'
to
6 1/2'

NOTE BELOW

3' CLEAR SPACE TO PROPERTY LINE

SERVICE ENTRANCE

FINAL GRADE

SERVICE EQUIPMENT

WHERE THE BUILDING IS TOO LOW TO OBTAIN PROPER CLEARANCES IT IS RECOMMENDED THAT CUSTOMER INSTALL AN UNDERGROUND SERVICE LATERAL.

3-20. *Service entrance riser support on a low building. (NMP)*

NOTE 1.

SERVICE LATERAL CABLE TO BE FURNISHED AND
INSTALLED BY CUSTOMER. IT MUST BE LONG
ENOUGH TO EXTEND 3 FEET ABOVE POINT "A"
WITHOUT A SPLICE. IF SECONDARY RACK IS NOT
ON POLE ASK COMPANY FOR ITS LOCATION.
PENDING CONNECTION BY COMPANY, TEMPORARILY
SECURE CABLE TO POLE TO PREVENT DAMAGE.

NOTE 2.

RISER CONDUIT SHALL BE GROUNDED EITHER BY
CONTINUITY OF GALVANIZED METALLIC CONDUIT TO THE
METER SOCKET TROUGH, OR BY BONDING THE POLE RISER
CONDUIT TO THE GROUNDING STUD IN THE METER SOCKET
USING AN INSULATED CONDUCTOR BURIED IN THE GROUND
EXTENDING UP THRU THE METER RISER CONDUIT, OR BY
BONDING TO THE GROUND NEUTRAL CONDUCTOR OF
THE SERVICE LATERAL.
RESISTANT BEND IN LOCATION SUBJECT TO HIGHWAY SALTING
USE CORROSION

45 DEGREES

PLAN AT A-A

A

COMPANY'S
SECONDARY
RACK
POINT "A"

SERVICE LATERAL CABLE
SEE NOTE 1

PIPE
STRAPS

RISER CONDUIT FURNISHED AND
INSTALLED BY CUSTOMER
CONSULT COMPANY FOR PROPER
LOCATION ON POLE

GROUND RISER CONDUIT
SEE NOTE 2

WHERE CONDUIT IS NOT CONTINUOUS USE INSULATING BUSHINGS
ON THESE ENDS OF CONDUIT AND BEND.

CONSTRUCTION FROM BUILDING TO POLE

8'
to
11'

ABOUT 3'

PROVIDE VERTICAL SURFACE FOR
MOUNTING METER SOCKET TROUGH

METER SOCKET TROUGH FURNISHED BY
COMPANY - INSTALLED BY CUSTOMER.

FACING METER

4'
to
6 1/2'

18" MIN. DEPTH

10'

3-21. *Underground service lateral from an overhead line service below 600 volts.* (NMP)

3-22. *Two-to six-meter installation, socket type. Single-phase, three-wire, 120/240 volts, 150-am-pere minimum service entrance. (NMP)*

vice. Notice the distance the meters should be above the floor. These are the plug-in type meter sockets. Note also the materials furnished by the power company and installed by the electrician. The meter board is usually painted black.

Mobile Home Installation

Mobile homes have been getting an increasing amount of attention from the National Electrical Code and other agencies. They are becoming a *permanent* dwelling for many people. These homes have particular needs that must be met to keep them electrically safe. Figure 3-24 shows a typical installation for mobile home parks.

BUILDING
WALL

INSIDE ←   → OUTSIDE

**SERVICE ENTRANCE
CONDUCTORS**

LEAVE LONG ENOUGH FOR
CONNECTION TO SERVICE
DROP BY COMPANY (24'')

WEATHERHEAD

TERMINATE CONDUIT ABOVE
SERVICE BRACKET

SECURELY ANCHOR SERVICE
BRACKET TO STRUCTURAL
FRAME MEMBERS OF WALL

SERVICE DROP

BELOW 400A– 1500 LBS. STRAIN
AT   400A–2000 LBS. STRAIN

WALL MUST BE
DESIGNED TO
WITHSTAND STRAINS
IMPOSED BY SERVICE
DROP

SERVICE BRACKET
FURNISHED BY COMPANY
INSTALLED BY CUSTOMER

GALVANIZED RIGID STEEL OR
ALUMINUM CONDUIT

STRAPS AT NOT MORE THAN
4 FOOT INTERVALS

AT LEAST 6"FROM
TOP OF METER BOARD
TO CEILING OR BEAM

150A
S.E.

METER BOARD
3/4"PLYWOOD, PAINTED
AND SECURELY MOUNTED

100A MINIMUM
METERED CUSTOMER
SERVICE EQUIPMENT

MAIN
SERVICE
EQUIPMENT

**GROUNDED
CONDUCTOR**

CONNECT TO
SERVICE EQUIPMENT
GROUND BLOCK

JUNCTION
BOX

UP TO SIX MAIN
SERVICE EQUIPMENTS
MAY BE INSTALLED

4 1/2'

COMPANY APPROVED
MULTIPLE METERING UNITS

INSTALL GROUNDING CONDUCTOR
IN APPROVED MANNER

FLOOR

3-23. *A typical indoor installation. More than six meters, socket type, installed for an apartment
complex. (NMP)*

CUSTOMER'S METER POLE

CONSULT COMPANY FOR POLE AND GUY REQUIREMENTS. POLE TO BE 5" MINIMUM DIAMETER AT TOP, 8" MINIMUM DIAMETER. 6 FT. FROM BUTT. NORMALLY 35 FT. POLE EXCEPT AS OTHERWISE NEEDED FOR REQUIRED SERVICE DROP HEIGHT. POLE TO BE TREATED WITH PENTACHLOROPHENOL IN OIL. EEI SPEC. TD100 OR ACCEPTABLE EQUIVALENT.

OVERHEAD FEEDER (ALTERNATE)

POLARIZED RECEPTACLE RECOMMENDED WITH GROUNDING POLE FOR CONNECTING ALL NON-CURRENT CARRYING METAL PARTS OF A TRAILER TO GROUND.

METER SOCKET TROUGHS
SUPPLIED BY COMPANY INSTALLED BY CUSTOMER

4 FEET CLEARANCE SHOULD BE MAINTAINED IN FRONT OF METERS AND SERVICE EQUIPMENT

OUTDOOR SERVICE EQUIPMENT 100 AMPERE MINIMUM, 3 WIRE, 125/250 VOLTS 150 AMPERE RECOMMENDED

EXTERIOR 3/4" PLYWOOD PANEL, PAINTED

WOOD POST PENTA TREATMENT SAME AS FOR POLE

UNDERGROUND FEEDER

6' MINIMUM

5/8" × 8' GROUND ROD

3' MINIMUM

4' TO 6 1/2'

SERVICE DROP

DIVISION OF OWNERSHIP

COMPANY — CUSTOMER

COMPANY'S SECONDARY

SERVICE BRACKET
SUPPLIED BY COMPANY INSTALLED BY CUSTOMER.

SERVICE ENTRANCE CABLE
OR SERVICE ENTRANCE CONDUCTORS IN GALVANIZED RIGID STEEL OR ALUMINUM CONDUIT

MINIMUM OF 4 METERS TO BE GROUPED AT EACH POINT. METERS CAN BE MOUNTED ON BOTH SIDES OF PANEL.

GROUNDING CONDUCTOR
#4 COPPER MINIMUM PROTECT WITH WOOD SHEATHING FOR ITS FULL LENGTH ON POLE

NOTES:
THIS SERVICE ARRANGEMENT IS TO BE USED FOR MOBILE HOME PARKS ONLY WHERE SINGLE METER OR GROUPED INDOOR METER INSTALLATION IS NOT PRACTICAL.

THE CUSTOMER SHOULD CONSULT COMPANY IN LOCATING SERVICE POLES AND GUYS AND BEFORE PROCEEDING WITH THIS TYPE OF INSTALLATION.

3-24. A typical installation for a mobile home park. (NMP)

QUESTIONS

1. Why are transformers needed in a distribution system?
2. What is a shareholder-owned electric company?
3. What are the advantages and disadvantages of an underground electrical distribution system?
4. What is the REA? When did it start? Why?
5. What is the phase of the power used in rural distribution lines?
6. What voltages are used in rural distribution lines?
7. Why are more communities demanding underground cables for service from the power company to the house?
8. How do service installations for mobile homes differ from those of permanent homes?

CHAPTER 4

Electrical Safety

Safety on the job is very important for the electrician. A number of safety procedures are needed to make certain that everyone on a construction site is protected. A building under construction can be an unsafe place. Thus, it is important that certain minimum conditions be met in terms of electrical wiring.

TEMPORARY WIRING

A temporary electrical wiring system does not have to be made with the detail and relative permanence needed for permanent wiring systems. However, the National Electrical Code (NEC) deals with this type of installation in Article 305. The article deals with the only ways a temporary wiring system may differ from a permanent system.

✓The NEC says that temporary electrical power and lighting installations shall be permitted during the construction, remodeling, or demolition of buildings, structures, or equipment.

✓Temporary electrical power and lighting installations are permitted during a period not to exceed ninety days for Christmas decorative lighting, carnivals, and experimental or development work.

Temporary feeders may be used. They should be supported by insulators spaced not over 10 feet apart if they are open conductors. They should terminate in a suitable or approved power outlet. This means one that is made by the manufacturer for job-site temporary wiring. Several variations of protection may be provided by portable receptable boxes. Fig. 4-1. These portable units have GFCI (ground fault circuit interrupters) located within them. They protect the entire circuit. A number of different cord sets are available for use with GFCI-protected plug-in units.

Most residential builders use a nonmetallic feeder and run it to a nonprotected receptacle. This does not provide adequate protection for the power tool user. Other parts of this chapter will show how the shorts in a power tool can cause serious shock hazards.

A responsible type of protection is preferred. It will pay for itself in a few jobs inasmuch as it is built to last. It does not have to be "jerry-rigged" each time a new construction job is undertaken.

TEMPORARY LIGHTING

✓Article 305-2 (Part f) of the NEC states that all lamps for general illumination shall be protected from accidental contact or breakage. Fig. 4-2. Protection shall be provided by elevation of at least 7 feet from normal working surface or by a suitable fixture or lampholder with a guard.

50A, 240/120 V
receptacle outlet

20A, 120 V
receptacles

50A, 240/120 V
outlets

50A, 240/120 V
4-conductor prewired
interconnecting cable

Receptacle
box

4-1. *Temporary branch circuits are manufactured for use at construction sites.*

4-2. *Temporary lighting strings of cable and sockets are available manufactured.*

Temporary wiring for new construction

Box not required
at splices and taps
in multiconductor
cords or cables or
in open wiring

4-3. *Splices made without boxes for temporary wiring of lights at a construction site.*

National Electrical Code accepts a 7 ft
mounting height as adequate protection without
a guard.

4-4. *Special watertight plugs and connectors are suggested for use at construction sites.*

4-5. *Three types of ladders.*

STRAIGHT

EXTENSION

FOLDING

VERTICAL HEIGHT

1/4 VERTICAL HEIGHT

4-6. *Proper placement of an extension ladder.*

Figure 4-3 shows how temporary wiring may be installed for use at a construction site. Splices may be made without boxes for the cord and cable runs.

TEMPORARY CORDS

Much information is available on extension cords that can be used on a construction site. More will be presented later in the chapter. Special water-tight plugs and connectors provide insurance against nuisance tripping of the GFCI. Fig. 4-4.

LADDERS

It is necessary for an electrician to work with outlets and fixtures that are above his or her reach. This requires a ladder. There are three types of ladders: folding, straight, and extension. Fig. 4-5.

Metal and water conduct electricity. Do not use a metal, metal-reinforced, or wet ladder where direct contact is possible with a power source.

For safety on a ladder, the pitch or angle should be such that the horizontal distance at the bottom is one-fourth the working length of the ladder. Fig. 4-6. Keep in

4-8. *Storage of an extension ladder.*

4-9. *The UL seal for a ladder.*

4-7. *Walking an extension ladder up.*

mind that the extension ladder should be placed so that the top of the ladder extends above the edge of the roof at least 3 feet.

To raise an extension-type ladder walk forward under the ladder as shown in Fig. 4-7. Move your hands to grasp other rungs as you proceed forward.

When you have finished, fold the ladder. Place it where it is not exposed to the weather or excess heat. To prevent sagging, an extension ladder should be stored horizontally on supports. Fig. 4-8.

Underwriters' Laboratories (UL) tests ladders. A label bearing their seal is placed on tested ladders. Fig. 4-9. This seal warns that metal ladders should not be used where contact may be made with electrical circuits. Some electrical inspectors require that the all metal ladders used on a jobsite have this seal.

In some residential construction, high ceilings make it necessary to construct scaffolding to reach some work areas. Figure 4-10 shows a few applications for such scaffolding. The scaffolding should be constructed so that it is absolutely secure. Makeshift scaffolding has been the cause of many accidents. It should be maneuverable and capable of being locked in position. Remove the scaffolding as soon as its use has been served. Store it in a safe location so it will not block any passageway.

There are a number of devices designed to protect people from their carelessness and from electrical shock. If you work around electricity for long you will probably receive a shock; most likely it will be caused by your own negligence. Such a shock could be fatal. There is very little difference between the experience of a slight tingle and a fatal shock. Eight milliamperes is 0.008 amperes and 50 milliamperes is 0.050 amperes. There is then, only a 0.042-ampere difference between the shock you feel as a tingle and the amount of current that will *kill* you. ✓

There have been cases where people have been shocked or burned by high voltages. Voltage does not necessarily kill — it is the *amount of current* through the body that does the damage.

These facts and figures are presented, not to scare you, but to make you aware of the dangers of being careless with the power in wires. This chapter will deal with some of the aspects of safety in electrical work. There are ways to be safe and live a long life. There are careless practices that are bound to mean a shorter life. Read this safety section with the idea that the life saved could easily be your own.

4-10A. *Adjustable scaffolding may be used by an electrician on the job.*

4-10B. *Manufactured scaffolding may be made with a platform to support the electrician.*

FATAL CURRENT

Strange as it seems, most fatal electric shocks happen to people who should know better. Here are some "electro-medical" facts that should make you think twice before taking chances.

It's the Current That Kills

Offhand, it would seem that a shock of 10 000 volts would be more deadly than a shock of 100 volts. But this is not so. Individuals have been electrocuted by appliances using ordinary house current of 110 volts and by electrical apparatus in industry using as little as 42 volts, direct current (DC). The real measure of a shock's intensity lies in the amount of current (milliamperes) forced through the body, and not in the voltage. Any electrical device used on a house wiring circuit can, under certain conditions, transmit a fatal current.

Any amount of current over 10 milliamperes (0.01 A) is capable of producing painful to severe shock. However, currents between 100 and 200 mA (0.1 to 0.2 A) are lethal.

From a practical viewpoint, after a person is knocked out by electric shock, it is impossible to tell how much current passed through the vital organs of the body. Artificial respiration must be applied immediately if breathing has stopped.

THE PHYSIOLOGICAL EFFECTS OF ELECTRIC SHOCK. Table 4-A shows the physiological effects of various current densities. Note that voltage is not a consideration. It takes voltage to make current flow. However, the amount of shock-current will vary, depending on the body resistance between the points of contact.

Table 4-A. Physiological Effects of Electric Currents

Readings		Effects
Safe Current Values	1 mA or less	Causes no sensation – not felt.
	1 mA to 8 mA	Sensation of shock, not painful; individual can let go at will since muscular control is not lost.
Unsafe Current Values	8 to 15 mA	Painful shock: individual can let go at will since muscular control is not lost.
	15 to 20 mA	Painful shock: control of adjacent muscles lost; victim cannot let go.
	20 to 50 mA	Painful, severe muscular contractions; breathing difficult.
	50 to 100 mA	Ventricular fibrillation – a heart condition that can result in instant death – is *possible*.
	100 to 200 mA	Ventricular fibrillation occurs.
	200 and over mA	Severe burns, severe muscular contractions – so severe that chest muscles clamp the heart and stop it for the duration of the shock. (This prevents ventricular fibrillation.)

*Information provided by National Safety Council.

As shown in Table 4-A, shock is more severe as the current rises. At values as low as 20 mA, breathing becomes labored, finally ceasing completely even at values below 75 mA.

As current approaches 100 mA, ventricular fibrillation of the heart occurs—an uncoordinated twitching of the walls of the heart's ventricles.

Note: A heart that is in fibrillation cannot be restricted by closed-chest cardiac massage. A special device called a defibrillator is available in some medical facilities and by ambulance services.

With more than 200 mA, muscular contractions are so severe that the heart is forcibly clamped during the shock. This clamping prevents the heart from going into ventricular fibrillation, making the victim's chances for survival better.

DANGER—LOW VOLTAGE. It is common knowledge that victims of high-voltage shock usually respond to artificial respiration more readily than the victims of low-voltage shock. The reason may be the above-mentioned clamping of the heart, due to the high current densities associated with high voltages. However, to prevent a misinterpretation of those details, remember that 75 volts are just as lethal as 750 volts.

The actual resistance of the body varies, depending upon the points of contact, and whether the skin is moist or dry. The area from one ear to the other, for example, has an internal resistance (which is lower than skin resistance) of only 100 ohms; from hand to foot it is nearer 500 ohms. Skin resistance may vary from 1000 ohms for wet skin, to more than 500 000 ohms for dry skin.

GENERAL SAFETY PRECAUTIONS

- *When working around electrical equipment, move slowly.* Make sure your feet are firmly placed for good balance. Don't lunge after falling tools. Kill all power and ground all high-voltage points before touching wiring. Make sure that power cannot be accidentally restored. Do not work on ungrounded equipment.
- *Do not examine live equipment when you are mentally or physically fatigued.* Keep one hand in your pocket while investigating live electrical equipment. Most important, do not touch electrical equipment while standing on metal floors, damp concrete, or other well-grounded surfaces. Do not handle electrical equipment while wearing damp clothing (particularly wet shoes), or while skin surfaces are damp.
- *Do not work alone!* Remember, the more you know about electrical equipment, the more heedless you're apt to become. Do not take unnecessary risks.

WHAT TO DO FOR VICTIMS

- *Cut voltage and/or remove victim from contact as quickly as possible—but without endangering your own safety.* Use a length of dry wood, rope, blanket, or similar device to pry or pull the victim loose. Don't waste valuable time looking for the power switch. The resistance of the victim's contact decreases with the passage of time. The fatal 100 mA to 200 mA level may be reached if action is delayed.
- *Start artificial respiration at once if the victim is unconscious and has stopped breathing.* Do not stop resuscitation until a medical authority pronounces the victim beyond help. It may take as long as eight hours to revive a patient. There may be no pulse. A condition similar to rigor mortis may be present. However, these are the manifestations of shock, not necessarily indications that the victim has died.

Table 4-B. Human Resistance to Electric Current

Body Area	Resistance (ohms)
Dry skin	100 000 to 600 000
Wet skin	1000
Internal body—	
hand to foot	400 to 600
Ear to ear	(about) 100

Table 4-C. Safe Distances from Live Circuits in Air

Do not approach live conductors closer than the following distances:

751 to 3 500 V	1 ft
3 501 to 10 000 V	2 ft
10 001 to 50 000 V	3 ft
50 001 to 100 000 V	5 ft
100 001 to 250 000 V	10 ft

Table 4-D. Flexible Cord Ampacities**

Size	Type S, SO, ST	Type SJ, SJO, SJT
AWG	Amperes	
18	7, 10*	7, 10*
16	10, 13*	10, 13*
14	15, 18*	
12	20	
10	25	
8	35	
6	45	
4	60	

**Table 400.5. Reprinted by permission from NFPA 70-1981, National Electrical Code, © 1980, National Fire Protection Association, Boston, Massachusetts.
*Where third conductor is used for equipment grounding only and does not carry load current.

Table 4-B gives some of the resistances the human body presents to an applied voltage. If you want to know the amount of current that would pass through the body, use Ohm's Law formulas and the applied voltage. Will it be enough to produce a fatal shock?

$$\text{Current through body} = \frac{\text{Voltage applied to body}}{\text{Resistance of body and contacts}}$$

Shock severity depends upon the following factors:
- Becomes more severe with increased voltage.
- Increases with the amount of moisture on contact surfaces.
- Increases with an increase in pressure of contact.
- Increases with an increase in area of body contact.
- Resistance of body portions.

The following three factors are involved in electric shock:
- Voltage (analogous to pressure forcing water in pipe).
- Current (analogous to water flowing in pipe).
- Resistance (analogous to something in pipe tending to hold back water flow).

Practical Problem

A worker with wet clothes, or wet with perspiration, comes in contact with a defective 120-volt light cord and establishes a good ground.

From preceding table:

Wet-skin resistance	1000 ohms
Body resistance	500 ohms
Internal resistance	1500 ohms

From formula:

$$\text{Current through body} = \frac{120 \text{ volts}}{1500 \text{ ohms}} = 80 \text{ mA}$$

Note that muscular control is lost and the victim cannot break contact. As the contact is continued, the skin resistance is reduced. If the skin is punctured, skin resistance may then be disregarded. Then for practical purposes, the total resistance of the worker may be in the neighborhood of 600 ohms.

$$\text{Current through body} = \frac{120 \text{ volts}}{600 \text{ ohms}} = 200 \text{ mA (certain death)}$$

There is always a danger of an electrical wire with high voltages leaking to ground through another path (if the ground wire is broken). You may become the path of least resistance if you come too close. Table 4-C.

To be safe, keep away from high-voltage wires. However, if you work with them or on them, make sure you use the proper safety procedures and clothing.

TYPICAL SHOCK HAZARDS WITH EXTENSION CORDS, PLUGS, AND RECEPTACLES

Extension Cords

A proper extension cord can make a difference in the safe operation of a piece of equipment. Make sure it has the proper capacity to handle the current needed. See Table 4-D.

Wire size and insulation of extension cords must be considered to assure proper operation with a piece of equipment. Take a look at Table 4-E for an explanation of the letters on a cord body. This table will help you determine the type of service for which a specific cord is recommended.

Table 4-E. Types of Flexible Cords

Trade name	Type letter	Size AWG	No. of conductors	Insulation braid on	Braid on each conductor	Outer covering	Use
Junior Hard Service Cord	SJ SJO SJT	18, 14	2, 3 or 4	Rubber Thermoplastic or Rubber	None	Rubber Oil Resist Compound Thermoplastic	Pendant or Portable Damp Places Hard Usage
Hard Service Cord	S SO ST	6 18, to 10 incl.	2 or more	Rubber Thermoplastic or Rubber	None	Rubber Oil Resist Compound Thermoplastic	Pendant or Portable Damp Places Hard Usage

Table 4-F. Extension Cord Sizes for Portable Electric Tools

THIS TABLE FOR 115-VOLT TOOLS						
Full-load ampere rating of tool	0 to 2.0 A	2.1 to 3.4 A	3.5 to 5.0 A	5.1 to 7.0 A	7.1 to 12.0 A	12.1 to 16.0 A
Length of Cord	Wire size (AWG)					
25 ft	18	18	18	16	14	14
50 ft	18	18	18	16	14	12
75 ft	18	18	16	14	12	10
100 ft	18	16	14	12	10	8
200 ft	16	14	12	10	8	6
300 ft	14	12	10	8	6	4
400 ft	12	10	8	6	4	4
500 ft	12	10	8	6	4	2
600 ft	10	8	6	4	2	2
800 ft	10	8	6	4	2	1
1000 ft	8	6	4	2	1	0

NOTE—If voltage is already low at the source (outlet), have voltage increased to standard, or use a much larger cable than listed in order to prevent any further loss in voltage.

Length of an extension cord is important in the sense that it must have the correct size wire to allow full voltage to reach the consuming device. For instance, a 25′ cord with No. 18 wire size is good for 2 amperes. If the distance is increased to 50 feet, you are still safe with No. 18 and 2 amperes. However, for a distance of 200 feet, the size of the wire must be increased to No. 16. This is necessary to carry the 2 amperes without dropping the voltage along the cord and therefore producing a low voltage at the consuming device at the end of the line. See Table 4-F.

INSPECTION AND PREVENTIVE MAINTENANCE. All drop or extension cords and other electrical equipment should be the proper size for the purpose intended. Wire size and insulation must be adequate for the service to be expected. Only equipment and cords listed by Underwriters' Laboratories, Inc. for the intended purpose should be used. Cords that are perfectly safe when purchased may be dangerous a short time later, unless they are given careful use. They should be kept free of oil and water. They should be protected from abrasion or other mechanical damage.

A system of preventive maintenance is suggested as a means of control over portable electrical equipment. Preventive maintenance involves inspection at regular intervals, the keeping of records of inspections, and making repairs when necessary. Splicing of extension cords should never be permitted.

A designated person, usually the electrician, should check the extension cords before they are put into use each day. The cord should have a continuous grounding conductor maintained throughout and carried back to the grounding bus at the service panel.

A visual inspection of the cord and harness should be made before each day's use. Check for loose or missing screws in the end and for loose or broken blades in the plug end. Fig. 4-11.

4-11. *Check the extension cord before it is placed in service each day.*

Check the continuity of each conductor with an ohmmeter at least once a week to make sure there are no broken conductors. Check the cord when it is first used or taken from the box. Check again after it has been repaired or after an incident such as shown in Fig. 4-12. If the plug has been repaired, check to see if the clamp on the wire is securely attached. Fig. 4-13.

Electric Plugs and Receptacles

Electric plugs and receptacles come in a wide variety of shapes and arrangements. Because there must be some regulation to insure that plugs and receptacles are used properly, the National Electrical Manufacturers Association (NEMA) has developed a standard for manufacturing these devices. Table 4-G will show how different voltages

4-12. *After an accident with construction equipment, the cord should be checked for damage.*

4-13. *Check the wire clamp on the plug for proper fit.*

Table 4-G. Voltage and Current Combinations for Plugs and Receptacles

Wiring Diagram	NEMA ANSI	Receptacle Configuration	Rating	
	5-15 C73.11		15A 125V	
	5-20 C73.12		20A 125V	
	5-30 C73.45		30A 125V	
	5-50 C73.46		50A 125V	
	6-15 C37.20		15A 250V	2 POLE 3 WIRE
	6-20 C73.51		20A 250V	
	6-30 C73.52		30A 250V	
	6-50 C73.53		50A 250V	
	7-15 C73.28		15A 277V	
	7-20 C73.63		20A 277V	
	7-30 C73.64		30A 277V	
	7-50 C73.65		50A 277V	
	10-20 C73.23		20A 125/250V	
	10-30 C73.24		30A 125/250V	3 POLE 3 WIRE
	10-50 C73.25		50A 125/250V	
	11-15 C73.54		15A 3φ 250V	
	11-20 C73.55		20A 3φ 250V	

	Rating	Receptacle Configuration	NEMA ANSI	Wiring Diagram
3P3W	30A 3φ 250V		11-30 C73.56	
	50A 3φ 250V		11-50 C73.57	
3 POLE 4 WIRE	15A 125/250V		14-15 C73.49	
	20A 125/250V		14-20 C73.50	
	30A 125/250V		14-30 C73.16	
	50A 125/250V		14-50 C73.17	
	60A 125/250V		14-60 C73.18	
	15A 3φ 250V		15-15 C73.58	
	20A 3φ 250V		15-20 C73.59	
	30A 3φ 250V		15-30 C73.60	
	50A 3φ 250V		15-50 C73.61	
	60A 3φ 250V		15-60 C73.62	
4 POLE 4 WIRE	15A 3φY 120/208V		18-15 C73.15	
	20A 3φY 120/208V		18-20 C73.26	
	30A 3φY 120/208V		18-30 C73.47	
	50A 3φY 120/208V		18-50 C73.48	
	60A 3φY 120/208V		18-60 C73.27	

Table 4-G (Continued)

	Rating	Receptacle Configuration	NEMA ANSI	Wiring Diagram
3P3W	30A 3φ 480V		L12-30 C73.102	
	30A 3φ 600V		L13-30 C73.103	
3 POLE 4 WIRE	20A 125/250V		L14-20 C73.83	
	30A 125/250V		L14-30 C73.84	
	20A 3φ 250V		L15-20 C73.85	
	30A 3φ 250V		L15-30 C73.86	
	20A 3φ 480V		L16-20 C73.87	
	30A 3φ 480V		L16-30 C73.88	
	30A 3φ 600V		L17-30 C73.89	
4 POLE 4 WIRE	20A 3φY 120/208V		L18-20 C73.104	
	30A 3φY 120/208V		L18-30 C73.105	
	20A 3φY 277/480V		L19-20 C73.106	
	30A 3φY 277/480V		L19-30 C73.107	
	20A 3φY 347/600V		L20-20 C73.108	
	30A 3φY 347/600V		L20-30 C73.109	
4P5W	20A 3φY 120/208V		L21-20 C73.90	
	30A 3φY 120/208V		L21-30 C73.91	

Wiring Diagram	NEMA ANSI	Receptacle Configuration	Rating	
	ML2 C73.44		15A 125V	**2 POLE 3 WIRE**
	L5-15 C73.42		15A 125V	
	L5-20 C73.72		20A 125V	
	L5-30 C73.73		30A 125V	
	L6-15 C73.74		15A 250V	
	L6-20 C73.75		20A 250V	
	L6-30 C73.76		30A 250V	
	L7-15 C73.43		15A 277V	
	L7-20 C73.77		20A 277V	
	L7-30 C73.78		30A 277V	
	L8-20 C73.79		20A 480V	
	L8-30 C73.80		30A 480V	
	L9-20 C73.81		20A 600V	
	L9-30 C73.82		30A 600V	
	ML3 C73.30		15A 125/250V	**3P3W**
	L10-20 C73.96		20A 125/250V	
	L10-30 C37.97		30A 125/250V	

Table 4-G (Continued)

Rating	Receptacle Configuration	NEMA ANSI	Wiring Diagram		Wiring Diagram	NEMA ANSI	Receptacle Configuration	Rating	
20A 3φY 277/480V		L22-20 C73.92				L11-15 C73.98		15A 3φ 250V	
30A 3φY 277/480V		L22-30 C73.93				L11-20 C73.99		20A 3φ 250V	
20A 3φY 347/600V		L23-20 C73.94				L11-30 C73.100		30A 3φ 250V	
30A 3φY 347/600V		L23-30 C73.95				L12-20 C73.101		20A 3φ 480V	

4 POLE 5 WIRE (left section)   *3 POLE 3 WIRE* (right section)

Bryant Electric Co.

EQUIPMENT GROUNDING

FUSE INTACT

HOT WIRE

SYSTEM GROUND

<u>THIS SYSTEM IS DANGEROUS</u>

DON'T YOU ACT AS A GROUNDING CONDUCTOR
– MAKE SURE EQUIPMENT IS GROUNDED THROUGH THE CORD!

4-14. *White-jacketed system grounds cannot conduct electricity from short circuits to the ground. Thus, they do not prevent the housing of faulty equipment from becoming charged. Therefore, a person who contacts the charged housing becomes the conductor in a short circuit to the ground. (National Safety Council)*

ELECTRICAL FAULT (SHORT)

DANGER: HOT!

SYSTEM GROUND

FUSE INTACT

HOT WIRE

FUSE BLOWS

HOT WIRE

SYSTEM GROUND

EQUIPMENT GROUNDING

THIS SYSTEM IS SAFE

LET THE GREEN GROUNDING CONDUCTOR TAKE THE CHARGE – NOT YOU!

ELECTRICAL FAULT (SHORT)

SAFE

SYSTEM GROUND

FUSE BLOWS

HOT WIRE

EQUIPMENT GROUNDING

4-15. *Equipment stays at ground potential in spite of the short circuit, if the circuit has a grounding conductor. Internal electrical faults cause current to short-circuit harmlessly to equipment ground in systems with green-jacketed grounding conductors. (National Safety Council)*

and current combinations are specified for a particular plug or receptacle. Note the wiring diagram, shown for three-phase, single-phase, wye and delta connections, and wye and delta with various grounds. You have to be very careful when wiring these plugs to make sure the proper terminal is connected.

Do not use a 3Ø plug or receptacle for 240-volt, three-wire, single-phase power.

• *A grounded conductor* has a white-colored jacket in a two- or three-wire cable (neutral wire). It is terminated to the white, or silver-colored terminal in a plug cap or connector; and it is terminated at the neutral bar in the distribution box.

An electrical fault may allow the hot line to contact the metal housing of electrical equipment, (in a typical two-wire system) or some other ungrounded conductor. In these cases, any person who touches that equipment or conductor will receive a shock. The person completes the circuit from the hot line to the ground, and current passes through the body. Because a body is not a good conductor, the current is not high enough to blow the fuse. It continues to pass through the body as long as the body remains in contact with the equipment. Fig. 4-14.

• *A grounding conductor,* or equipment ground, is a wire attached to the housing, or other conductive parts of electrical equipment that are not normally energized, to carry current from them to the ground. Thus, if a person touches a part that is accidentally energized, there will be no shock, because the grounding line furnishes a much lower resistance path to the ground. Fig. 4-15. Moreover, the high current pass-

4-16. A 125-volt, 15-ampere standard parallel configuration. (National Safety Council)

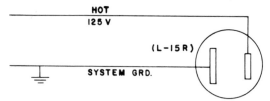

4-17A. A 125-volt, 15-ampere polarized parallel configuration. (National Safety Council)

4-17B. A 125-volt, 15-ampere polarized locking configuration. (National Safety Council)

4-18. A 250-volt, 15-ampere tandem configuration. (National Safety Council)

4-19. NEMA 125-volt, 15-ampere U-blade configuration. (National Safety Council)

ing through the wire conductor blows the fuse and stops the current. In normal operation, a grounding conductor does not carry current.

The grounding conductor in a three-wire conductor cable has a green jacket. It is always terminated at the green-colored hexhead screw on the cap or connector. It utilizes either a green-colored conductor or a metallic conduit as its path to ground. In Canada, this conductor is referred to as the *earthing conductor.* This term is somewhat more descriptive and helpful in distinguishing between grounding conductors and neutral wires, or grounded conductors.

• *Two-wire configurations.* The standard parallel configuration, illustrated in Fig. 4-16, is designated a 125-volt, 15-ampere, two-pole, two-wire type at the present time. It will be phased out as old wiring is replaced with newer, two-wire with ground plugs and receptacles. These older types are no longer permitted by the National Electrical Code.

A polarized parallel configuration, Fig. 4-17, is also designated a 125-volt, 15-ampere, two-pole, two-wire type. It differs from the standard parallel only in that one of the blades is about $\frac{1}{16}''$ wider than the other. This wider blade is intended for connection to the neutral wire. Cord connectors and receptacles in the configuration also have one wide slot. This assures that, when properly wired, the neutral conductor will be continuous throughout the entire system. Connectors have one wide slot, enabling them to accommodate both the standard and polarized parallel caps. This type of plug may be noticed on some previous models of television sets, where a "hot chassis" was used, making it important to connect the ground wire to the chassis.

Another old type of plug (cap) you may see in some locations is the tandem-position blades (now banned). Fig. 4-18. This is a 250-volt, 15-ampere, two-pole, two-wire type that is used for 230-volt, single-phase circuits where there is no separate equipment ground. The most frequent applications will be found on fractional-horsepower, single-phase motors with 230 volts, where a separate equipment ground in the supply cord is not required. In some cases, grounding is accomplished by a separate flexible wire or strap.

● *Three-wire configurations.* The U-blade configuration, Fig. 4-19, is most widely used and is designated a 125-volt, 15-ampere, two-pole, three-wire, grounding type. The grounding blade on the plug is slightly longer than the two line blades. The configuration is used for 115-volt, single-phase applications that require a separate equipment ground. It is growing in popularity. It will probably replace the two-pole, two-wire configuration as the older type is phased out by replacements.

● *System ground.* A system ground is the grounding of the neutral conductor, or ground leg, of the circuit to prevent lightning or other high voltages from exceeding the design limits of the circuit. It also limits the maximum potential to ground due to normal voltage.

● *Equipment ground.* An equipment ground is the grounding of exposed conductive materials, such as conduit, switch boxes, or meter frames, which enclose conductors and equipment. This is to prevent the equipment from exceeding ground potential.

GROUNDING ADAPTERS

One of the problems with a grounding adapter is the lack of a grounded screw to which to attach the green pigtail at the plug. In Fig. 4-20, the ground wire has been

4-20. *One method of connecting a ground wire to an ungrounded receptacle.*

GREEN WIRE

LUG

WALL RECEPTACLE

ADAPTER ASSEMBLY

ATTACH TO GROUNDED COLD WATER PIPE (Remove Paint, etc.)

GROUND WIRE (Minimum 18 Gauge)

POWER SUPPLY CORD (with Three-Prong Grounding Plug)

added and attached to a grounded cold water pipe. In some cases, there is no ground wire in the cable that services the outlet. Therefore, there is no ground to the metal box. In some instances, the box is an insulated-type made of nonmetallic materials. In other cases, the old knob-and-tube wiring may service the outlet or receptacle.

These are some of the reasons why the National Electrical Code (NEC) prohibits the use of such a pigtail adapter. They are, however, widely available and used. Refer to Fig. 4-21 to see why this could be a fatal mistake. Once the pigtail adapter contacts the hot side of the outlet — it has not been attached to the faceplate screw since it would take time to do so — the handle of the drill is *hot* with 120 volts. If the person holding the drill touches a ground anywhere, the path for current flow is through the body.

One plug type is recognized by both the UL and the NE Code. Fig. 4-22. The grounding adapter has a rigid tab with a spade end. This tab connects the adapter's grounding terminal to the metal screw contacting the grounded metal yoke that mounts the receptacle to the grounded metal outlet box. If a nonmetallic outlet box is used, the adapter's grounding terminal may be connected to an equipment grounding conductor in nonmetallic NM cable. Different-width blades on the adapter permit insertion in only one way into the polarized blade opening on the receptacle.

Special Ground-Fault Protective Devices

Fuses and circuit breakers cannot distinguish between fault current and normal load current. A ground-fault interrupter, however, is insensitive to load current. It responds only to a ground-fault current.

Dangling unconnected flexible grounding pigtail

Three-prong plug cap for cord-connected tool inserted into top half of nongrounding-type duplex receptacle

Green ground wire accidentally contacts 120-volt terminal

HOT CONTACT

Green

Black
White
To motor

Three wires in cord to tool

DANGER!!
120 VOLTS
FROM DRILL
CASE TO
GROUND

4-21. *A pigtail adapter can cause problems if not properly attached to the screw.*

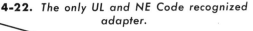

4-22. *The only UL and NE Code recognized adapter.*

4-23. *This is a typical residential ground using armored cable to positively ground the conductor from the service enclosure to the grounding electrode water pipe. (Thomas & Betts)*

4-24. *This clamp bonds the jumpers around meters or other breaks. No additional hubs are needed. (Thomas & Betts)*

An imbalance between a phase wire and neutral, or between phase wires of as little as 3 milliamperes, will cause a ground-fault interrupter to operate in less than 50 milliseconds. Portable and permanently installed devices are available. These will provide safe, portable-tool installation, independent of the ground return path.

The National Electrical Code recognizes the importance of a path to ground of continuous low impedance. No other problem is dealt with more thoroughly. The Code requires certain basic practices to be followed to make the equipment-grounding system safe. It recognizes the importance of Underwriters' Laboratories listings of various connectors used in a system.

Code Requirements for Grounding Conductors

The National Electrical Code requires that a system grounding conductor be connected to any local metallic water-piping system available on the premises, provided that the length of the buried water piping is a minimun of 10 feet. A system may be less than 10 feet long or its electrical continuity may be broken by either disconnection or nonmetallic fittings. In these cases, it should be supplemented by the use of an additional electrode of a type specified by Section 250-83. Figures 4-23 and 4-24 illustrate the various types of connectors for grounding or bonding.

4-25. Grounding bushings must maintain good electrical contact with the enclosure. The lay-in lug provides unusual flexibility in positioning and fastening the grounding conductor. (Thomas & Betts)

• *Fittings.* Whatever the type of grounding electrode, a system grounding conductor should be connected through any one of a number of UL-listed clamps. Grounding wire may be either bare, insulated, or armored conductor, or it may be in EMT (thinwall conduit) or rigid conduit. Selection of the proper fitting depends upon the type of conductor and conduit and the type and diameter of the grounding electrode.

An approved clamp should be made of cast bronze or brass, or of plain or malleable cast iron. In attaching the ground clamp, care should be taken to make sure that any nonconductive protective coating, such as paint or enamel, is removed. Also, the surfaces must be clean, to insure a good electrical connection.

• *Bonding jumpers.* Wherever electrical continuity of the grounding circuit may be interrupted, either by insulation or by disconnection, a bonding jumper is required by the Code. If a grounding attachment to metallic underground water piping is not made on the street side of a water meter, a bonding jumper must be used around the meter. This is needed to insure an uninterrupted path to ground. Bonding jumpers must also be used to span other items likely to become disconnected, such as valves and service unions. In addition, they must be used to span insulating sections, such as plastic pipe, certain water softeners, or other insulating links. Bonding jumpers require the same size conductor as that of the grounding conductor which is a run to water pipe, or other grounding electrode.

SERVICE EQUIPMENT. Continuity at the service equipment is essential and must be assured. The fittings shown in Figs. 4-25 and 4-26, when properly applied, will help assure the continuity of ground. Connections of the fittings are made at the metallic

4-26. This grounding wedge works equally well on existing or new installations. This design reduces installing time while assuring continuity of ground. (Thomas & Betts)

enclosures by means of grounding wedges, locknuts, and bushings. They are installed so as to insure continuity of ground. Thus, care must be exercised to select fittings that will maintain good contact. Bonding locknuts should be used if the raceway does not use the largest concentric knockout.

Grounding bushings may have an insulated throat to protect the insulation of the current-carrying wire. However, this does not affect their function as a grounding connection. Good electrical contact is maintained with the metallic enclosure by the ground wire attached to the screw. Good electrical contact is also maintained with the grounding conduit. All threaded fittings must be made up wrench-tight, as specified by the Code.

The system grounding conductor is attached to the white wire (the grounded circuit conductor) within the service equipment. It should run without joint or splice to the grounding electrode. Generally, attachment of the grounding conductor is made to a built-in terminal on the neutral bar. The neutral bar terminal may not accommodate the size of grounding conductor required. Then, a pressure terminal of the correct size must be bolted on and used for the system ground attachment.

• *Bonding.* To insure electrical continuity of the grounding circuit, bonding is required at all conduit connections in service equipment. It is required also at points where any nonconductive coating exists that might impair such continuity.

• *Where bonding is required.* Bonding is required at connections between service raceways, service cable armor, and at all service equipment enclosures containing service entrance conductors (including meter fittings and boxes). It is required also at any conduit or armor that forms part of the grounding conductor to the service raceway.

When meter housings are mounted outdoors, separate from the service equipment, the neutral wire in the meter should run to ground.

A meter located inside a building should have the neutral wire bonded to the meter housing and attached to the grounded service conductor.

On rigid conduit, good ground connections will be assured at the point where conduit meets the tapped hub or boss on the enclosure. If service-entrance cable is used, it is recommended that watertight connectors be installed, to protect the joint from moisture.

An iron-bodied, universal-type entrance cap is recommended where service conductors from supply enter the building. Such a fitting works equally well on rigid conduit or on thinwall conduit with no threading. These caps are of weatherproof design, and function to protect, as well as to fasten conductors to the structure. These caps are also available in aluminum.

An emphatic warning is repeated throughout the National Electrical Code: NEVER USE SOLDER FOR GROUNDING CONNECTIONS.

SAFETY DEVICES

As a power source, electricity can create conditions almost certain to result in bodily harm, property damage, or both. It is important for workers to understand the hazards involved when they are working around electrical power tools, maintaining electrical equipment, or installing equipment for electrical operation.

• *Overcurrent devices* should be installed in every circuit. They should be of the same size and type to interrupt current flow when it exceeds the capacity of the conductor. Proper selection takes into account not only the capacity of the conductor, but also the rating of the power supply and potential short circuits.

- *Fuses* must be chosen according to type and capacity to fill a specific need. It is recommended that a switch be placed in the circuit so that fuses can be deenergized before they are handled. Insulated fuse pullers should also be used. Blown fuses should be replaced by others of the same type. Fuses should never be inserted in a live circuit.

- *Circuit breakers* have long been used in high-voltage circuits with large current capacities. Recently they have become more common in other types of circuits. Circuit breakers should be selected for a specific installation by qualified engineers. They should be checked regularly by experienced maintenance personnel.

- A *ground-fault circuit interrupter* is a fast-operating circuit breaker that is sensitive to very low levels of current leakage to ground. The interrupter is designed to limit electric shock to a current- and time-duration value below that which can produce serious injury. The unit operates *only* on line-to-ground fault currents, such as insulation leakage currents, or currents likely to flow during accidental contact with a "hot" wire of a 120-volt circuit and ground. *It does not protect in the event of a line-to-line contact.* There are two main types of GFIs:

1. The differential ground-fault interrupter, available with various modifications, has current-carrying conductors passing through the circular iron core of a doughnut-shaped differential transformer. As long as all the electricity passes through the transformer, the differential transformer is not affected, and will not trigger the sensing circuit. If a portion of the current flows to ground and through the fault-detector line, however, the flow of electricity through the sensing windings of the differential transformer causes the sensing circuit to open the circuit breaker. These devices can be arranged to interrupt a circuit for currents of as little as 5 mA flowing to ground.

2. Another design is the isolation-type ground-fault interrupter. This unit combines the safety of an isolation system with the response of an electronic sensing circuit. In this setup, an isolating transformer provides an inductive coupling between load and line. Both hot and neutral wires connect to the isolating transformer. There is no continuous wire between.

In the latter type of interrupter, a ground fault must pass through the electronic sensing circuit, which has sufficient resistance to limit current flow to as low as 2 mA. This is well below the level of human perception.

The ground-fault interrupters by Hubbell operate on the differential transformer principle. They monitor load currents flowing in the protected circuit, comparing current flowing to the load with current flowing from it. Leakage to ground (ground fault) appears as the difference between these currents. When this exceeds a preset level, the GFI interrupts the circuit.

Figure 4-27 is a mini-portable plug-in type of GFI with a 120-volt, 15-ampere capability. Just plug it into the socket and then plug the device to be used on 120 volts and under 15 amperes into the GFI. It guards against shock hazard. It is used with pumps, extension cords and hand tools as well as many other applications in the home, office or factory. Safety minded electricians use this unit for on-the-job protection.

This device has a built-in test circuit that imposes an artificial ground fault on the load circuit to assure that the ground-fault protection is functioning properly. This particular device trips at 5 mA.

Some GFIs may be used in conjunction with underwater lights in older pools (built prior to 1965) which required less sensitive equipment, or one that has a 20 mA sensitivity. Some less sensitive types may also be used in industrial applications where moisture and chemicals preclude the use of 5 mA type.

Figure 4-28 shows the wiring of a GFI for use with a swimming pool. Pools require either a Class A interrupter, which calls for a sensitivity of 5 mA to ground, or a Class B model, which has a sensitivity of 20 mA to ground.

4-27. *Mini-portable, plug-in type ground fault circuit interrupter (GFI or GFCI) for use on 120-volt, 15-ampere circuits.*

4-28. *Wiring diagram for GFI designed for wet locations and outdoor installations. (Harvey Hubbell)*

A lightweight, compact GFI is available. It has four outlets. One automatically trips when the supply voltage is interrupted, or when the line voltage drops too low. It may be plugged into any 120-volt, 60-hertz line that has the proper over-current protection. The construction industry uses it where frequent damage to electrical equipment develops ground faults. Use of the portable GFI can help protect an employer by preventing liability cases or claims.

GFI-Related Code Requirements

This is a paraphrase summary of the various Sections of the 1978 National Electrical Code that relate to ground-fault circuit interrupters:

Section 210-8(a): This requires that all 15- and 20-ampere, single-phase, 120-volt receptacles installed outdoors and in bathrooms of *all* residential occupancies are to be protected with ground fault circuit interrupters. Fig. 4-29.

Section 210-8(b): This requirement makes it mandatory for all 15- and 20-ampere, 120-volt, single-phase receptacles on construction sites which are not part of the permanent wiring of the building or structure to be protected by GFIs for personnel. An exception to this basic requirement is that receptacles on portable generators 5 kW or less need not be protected. Figs. 4-30 and 4-31.

Section 210-52: This basic section tells you how many and where you have to put receptacle outlets. Of specific interest is the fact that it requires a receptacle outlet in

Duplex receptacle outdoors with weatherproof cover and grade-level access

Earth

Receptacles fed out of ground must be GFI protected

4-29. *Receptacles fed out of the ground must be GFI protected. Note locations of outdoor receptacles and those that require a GFI device.*

Grade-level patio or open porch with receptacle accessible from grade level

No grade-level access: no stairs to second-floor deck porch of one-family house—only access is door from house

Open-front porch of one-family house, with access by steps from grade level

Outdoor receptacle on second-floor deck porch would not require GFI protection, but receptacle on porch below would need GFI protection.

Outdoor receptacle on this porch might be used to connect appliance used on ground, and GFI protection would seem to be required.

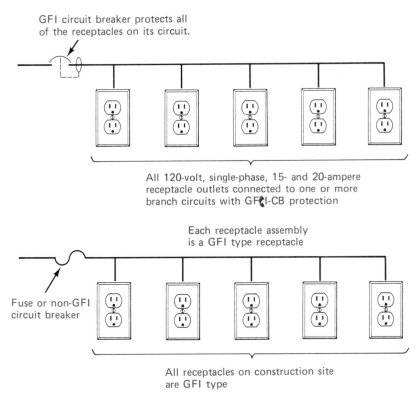

GFI circuit breaker protects all
of the receptacles on its circuit.

All 120-volt, single-phase, 15- and 20-ampere
receptacle outlets connected to one or more
branch circuits with GFCI-CB protection

Each receptacle assembly
is a GFI type receptacle

Fuse or non-GFI
circuit breaker

All receptacles on construction site
are GFI type

4-30. *Ways that satisfy the basic rule on personnel shock protection at construction sites.*

the bathroom *adjacent to the basin location.* Further, it requires that there be at least one receptacle outlet installed outdoors on all one-family dwellings. Finally, this section indicates that the receptacle outlets that are required by the section, such as that in the bathroom, must be in addition to any receptacles that are part of a lighting fixture or appliance or that are located within cabinets or cupboards.

Section 550-6(b): Article 550 deals with mobile homes. The fine print note following this section indicates that the ground-fault circuit interrupter requirements in Section 210-8(a) which apply to outdoor and bathroom receptacles in residences are also applicable to those receptacles in mobile homes.

Section 550-3(a): This basic requirement concerns the receptacles in mobile home service equipment (pedestals). It indicates that, where receptacles are required in this power outlet to supply the mobile home, they shall be 50-ampere receptacles. If there are additional receptacles in this power outlet of the 15- or 20-ampere, 120-volt, single-phase type, these additional receptacles shall be protected with GFIs.

Section 680-6(a): Article 680 deals with swimming pools. This particular requirement says that no receptacle on the property shall be located within 10 feet of the inside walls of the pool, and that any receptacle located 10 to 15 feet from the inside walls shall be protected with a GFI.

Fuse or non-GFI
circuit breaker . . .

. . . And all receptacles
are non-GFI type

BUT. . .when any receptacle
is PUT INTO USE—that is, an
employee connects a tool or
appliance to the receptacle, a
portable GFI assembly is
inserted into the conventional
receptacle. The cord cap is
plugged into the GFI device
(or a cord-connected GFI
unit could be inserted).

Plug blade
assembly
on back of
portable
GFI device

Portable cord
or tool cord
plugs in here

These 15-ampere receptacles are fed by a
temporary branch circuit without ground-
fault protection ahead of them.

To other
receptacles

Temporary
panel

Wherever personnel are using cord-connected
tools they plug in this portable ground-fault
circuit interrupter having protected receptacles
on its face for connection of the tools.

4-31. *Portable GFI devices and how they satisfy the NE Code requirements.*

Section 680-6(b)(1): This section pertains to lighting fixtures installed around swimming pools. It requires that any lighting fixture 5 feet horizontally from the inside walls of the pool be protected by a GFI unless 12 feet above the maximum water level.

Section 680-6(b)(2): When a new swimming pool is installed next to an existing building with a lighting fixture on the building, it's possible for that existing lighting fixture to be within 5 feet of the inside edge of the pool. If this is the case, then the existing lighting fixture must be protected by a GFI.

Section 680-6(b)(3): This section requires GFI protection of any lighting fixture where the fixture or its supporting means is located within 16 feet of any point on the water surface.

Section 680-20(a)(1): This requirement mandates the use of GFIs to protect all electrical equipment, including the power supply cords used with storable (above-ground) swimming pools.

Section 680-50: This requirement applies to fountains. It requires that all electrical equipment such as lighting fixtures, submersible pumps, etc., be protected by GFIs except where that equipment is operated at 15 volts or less.

QUESTIONS

1. What kills—current or voltage?
2. What are the physiological effects of electrical shock?
3. Why is low voltage considered dangerous in some cases?
4. How much resistance does the human body have when the skin is dry?
5. How can you determine the amount of current through your body when wet?
6. What are safe distances for operating around live circuits?
7. Why is it important to choose the right extension cord for the job?
8. Why are extension cords unsafe in some locations?
9. What is an electrical ground?
10. Why is improper grounding dangerous?
11. What are ten basic factors considered important in working safely around electricity?

CHAPTER 5

Wiring to the Electrical Code Requirements

Electricity has not been around too long in the relative span of time. Still, it is old enough to be considered an energy source no one would want to be without.

This was not always the case, however. Thomas Edison met with resistance from those who preferred gas lighting. To this day, we have some streets lighted by gas. But open flame, with its constant threat of uncontrolled fires, has always been a danger to people and their wooden buildings.

Some examples of the destruction that fire has wrought: Property worth more than $15 million was destroyed in New York City when 13 acres [5.2 hectares] burned in 1835. Much of the town of Charleston, South Carolina, burned to the ground in 1861, with $10 million damage. In Portland, Maine, following the July 4th celebration in 1866, a $10 million blaze destroyed 1500 buildings. At today's prices this would be billions instead of millions.

Because these fires meant heavy losses to fire insurance companies, they began a cooperative effort to *prevent* fires. The New York Board of Fire Underwriters was formed in May of 1867. The function of this organization was to arrange for inspection of premises. It was also to make recommendations for the removal of possible fire hazards. In 1882, the Committee of Surveys drew up a set of safeguards for *arc and incandescent lighting.* These safeguards were the forerunners of the present National Electrical Code.

A letter from Thomas Edison to the New York Board of Fire Underwriters, in May of 1881, shows how firmly he believed that his electric lighting company was " . . . free from any possible danger from fire." He believed his system of lighting was far superior to natural gas, that it provided protection from fire, and was safer in all regards.

As you know, electricity can cause shock and fire when improperly used or installed. This is why there are a number of codes written for the protection of those who generate and use electricity.

NATIONAL ELECTRICAL CODE (NEC)

The National Electrical Code was first developed in 1911. It is sponsored by the National Fire Protection Association, an organization established in 1896.

The Code is reprinted every three years. There are a number of changes each time a new edition is issued. Additions or changes are agreed upon by a committee which, after studying recommendations from many sources, passes on their final form and wording before presenting them to the Association for publishing.

"This *Code* is purely advisory as far as the National Fire Protection Association and National Standards Institute are concerned, but it is offered for use in law and for regulatory purposes in the interest of life and protection of property."

The National Electrical Code serves as a basis on which local governmental authorities can write ordinances that deal with protection of the lives of people working with or using electricity or electrical devices. Local laws almost always refer to the National Electrical Code as the "standard minimum," sometimes making additions to it to meet local conditions.

For example, burying of metal in some types of soils can cause deterioration of the metal used for grounding, or for conduits, more quickly than for most other locations. This unique soil condition must therefore be pointed out in a *local* modification to the National Electrical Code.

The Code is concerned with the *how* of electrical installation. It states the approved quantity and sizes of wire that can safely be used in a conduit as well as the number of wires in a box, and whether the wire should be aluminum or copper.

For example, No. 12 AWG aluminum wire, also copper-clad, can carry 15 amperes safely, if the types of wire are *RUW, T, TW* and if the temperature in which the wire is installed for operation does not exceed 60°C [140° F.]. This assumes that there are not more than three conductors in a raceway (conduit) or buried cable.

RUW means the insulation is made of moisture-resistant latex rubber and is suitable for dry and wet locations with a maximum operating temperature of 60°C [140° F.].

T means the insulation is thermoplastic, with the same maximum temperature, but for use in dry locations only.

TW means the insulation is moisture-resistant thermoplastic, for the same maximum temperatures, and can be used in both dry and wet locations.

The National Electrical Code may be bought in paperback or as a hardcover book. It contains much of the information needed by an electrician on the job. It has definitions of electrical terms and chapters on wiring design and protection; wiring methods and materials; equipment for general use; special occupancies, equipment, and conditions; sections on communications; and tables, an index, and an appendix. A membership list of the National Electrical Code (NEC) Committee is included, stating titles held, and addresses of individual members. The members come from all parts of the country. This gives greater input to Code revisions, keeping it current and, as nearly as possible, applicable to the entire country. The 1971 Code, for example, contains a very much revised article on swimming pools, fountains, spas or hot tubs, storable swimming pools, and wading pools (Article 680). The use of *ground-fault interrupters* is called for here to protect those around the pool as well as in the water. Grounding methods for electrical devices are stressed in the handbook.

No electrician should be without the latest edition of the Code book in his or her tool box. It has answers to most wiring problems. It may be obtained from:

National Fire Protection Association
470 Atlantic Avenue
Boston, Massachusetts 02210

UNDERWRITERS' LABORATORIES

Underwriters' Laboratories, Inc., was founded in 1894 by William Henry Merrill. He came to Chicago to test the installation of Thomas Edison's new incandescent electric light at the Columbian Exposition. He later started the UL for insurance companies to test products for electric and fire hazards. It continued as a testing laboratory for insurance underwriters until 1917. It then became an independent, self-supporting, safety-testing laboratory. The National Board of Fire Underwriters (now American Insurance Association) continued as sponsors of UL until 1968. At that time, sponsorship and membership were broadened to include representatives of consumer interests, governmental bodies or agencies, education, public safety bodies, public utilities, and the insurance industry, in addition to safety and standardization experts.

UL has expanded its testing services to more than 13 000 manufacturers throughout the world. Over one billion UL labels are used each year on products listed by Underwriters' Laboratories.

UL is chartered as a not-for-profit organization without capital stock, under the laws of the State of Delaware. Its function is to establish, maintain, and operate laboratories for the examination and testing of devices, systems, and materials.

Its objectives, as stated by UL, are: "By scientific investigation, study, experiments and tests, to determine the relation of various materials, devices, products, equipment, constructions, methods and systems to hazards appurtenant thereto or to the use thereof, affecting life and property, and to ascertain, define and publish standards, classifications and specifications for materials, devices, equipment, construction, methods and systems affecting such hazards, and other information tending to reduce and prevent loss of life and property from such hazards."

The corporate headquarters, together with one of the testing laboratories, is located at 207 East Ohio Street, Chicago, Illinois. Other offices and laboratories are located at Northbrook, Illinois; at Melville, New York; at Santa Clara, California; and at Tampa, Florida.

5-1. *Testing furnaces and air conditioners at the Underwriters' Laboratories. (UL)*

5-2. *Checking electrical equipment at UL. (UL)*

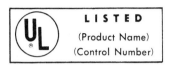

5-3. *Testing fire extinguishers at UL. (UL)*

5-4. *UL stickers and labels. (UL)*

Underwriters' Laboratories, Inc., has a total staff of over 2000 employees. More than 700 persons are engaged in engineering work. Of this number, approximately 425 are graduate engineers. Fig. 5-1. Supplementing the engineering staff are more than 500 factory inspectors.

Engineering functions of Underwriters' Laboratories are divided among these six departments:

- Burglary protection and signaling.
- Casualty and chemical hazards.
- Electricity. Fig. 5-2.
- Fire protection. Fig. 5-3.
- Heating, air conditioning, and refrigeration.
- Marine equipment.

The electrical is the largest of the six engineering departments. Safety evaluations are made on hundreds of different types of appliances for use in homes, commercial buildings, schools, and factories. The scope of the work in this department includes electrical construction product in detail, particularly its construction and test performance. The procedure becomes the manufacturer's guide for future production. It is used by the Laboratories' inspectors for periodic review of the listed product. Procedures are kept up to date by the manufacturer, who informs the Laboratories of any changes in the product or additions to the line. Fig. 5-4.

CANADIAN STANDARDS ASSOCIATION (CSA)

History of the CSA

The Canadian economy expanded after the turn of the century. It then became apparent that technical standardization was essential if Canada was to compete in the world markets. The original standards association was incorporated under the Dominion Companies Act in 1919, with the name Canadian Engineering Standards Association, a nonprofit, independent, voluntary organization.

By Supplementary Letters Patent, in 1944, this association became the Canadian Standards Association and extended its activities to a broader field of standardization.

Purpose of the Association

The basic objectives of the CSA are:
- To develop voluntary national standards.
- To provide certification services for national standards.
- To represent Canada in international standards activities.

Location and Operation of the Association

CSA headquarters and major testing facilities are located in Rexdale, a suburb of Toronto, in Ontario, Canada. These facilities house a staff of over 400, which includes professional engineers and skilled laboratory technicians, in addition to management and secretarial personnel.

Regional offices and test facilities in Montreal, Winnipeg, and Vancouver are maintained for the convenience of clients across Canada. There is a branch office in Edmonton, Alberta. Executive offices are located in the national capital of Ottawa, Ontario.

THE CSA MARK. Products certified by CSA are eligible to bear the CSA certification mark. Misuse of the mark may result in suspension or cancellation of certification. CSA may resort to legal action to protect its registered trademark in the event of abuse. Fig. 5-5. In addition, CSA information tags and other markings are also made available for certified products and their containers, to supplement the CSA mark.

5-5. *The CSA trademark. (CSA)*

For your own safety, the CSA symbol on electrical equipment you purchase is an assurance that such equipment has passed rigid inspection.

WIRING ACCORDING TO THE NATIONAL ELECTRICAL CODE

There are a number of methods of wiring. Some use materials totally unacceptable in certain areas of the United States. However, local codes vary. This means the National Electrical Code must establish standards for all the various types of wiring materials. Even the knob-and-tube method shown in Chapter 14, although obsolete, is still used in some areas of the United States. Therefore, the Code has established certain proper procedures in working with this type of wiring system. It can be added to, but it cannot be installed in any new construction

BX cable, which is armored cable, is specified in some locations. It is not permitted in other parts of the country. However, the Code has certain established rules that regulate its use.

5-6. Armored cable. Steel armor is wound around the wires for protection. Note the grounding strip.

GROUNDING STRIP

5-7. Cutting armored cable with a hacksaw.

Armored Cable

Type AC (armored cable) is familiarly used in all types of electrical systems for power and light branch circuits, as well as feeders.

Type AC is also used for signal and control circuit work. It is called for in some places because of its mechanical strength. The wires inside are protected by the metal covering. Fig. 5-6. The steel armor protects the wires. A bare grounding strip runs inside the armor for better equipment grounding.

WORKING WITH ARMORED CABLE (AC). Working with armored cable (AC) can be difficult. In sawing the armored cable, a hacksaw should be held as shown in Fig. 5-7. Once the armor has been cut with the hacksaw, a twist will remove the unwanted end and expose the wires needed for making a connection. See Fig. 5-8 for a different

5-8. Using a newer tool to do a cleaner job of cutting the armored cable. (Seatek)

5-9. *Twisting the cut end of armor to remove it.*

5-10. *Unwrapping the paper from the exposed end.*

5-11. *Snapping the paper to break it off inside the armored cable.*

ANTI-SHORT

5-12. *Inserting the fiber bushing for antishorting purposes.*

5-13. *The fiber bushing used to protect against damage to the wires in armored cable.*

5-14. *Placing a connector on armored cable.*

5-15. *Driving the locknut in place within the box to hold the connector of the armored cable securely.*

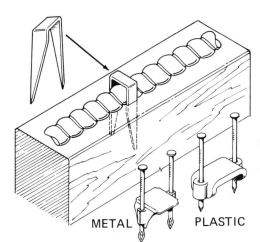

5-16. *Staples are used to support armored cable on wooden surfaces.*

METAL   PLASTIC

method of cutting armored cable. The Roto-Split® cutter provides a clean cut and saves time as compared with the hacksaw Figure 5-9 shows what happens once the cut cable is twisted.

After twisting you must unwrap the paper over the wires as far as you can. Do not tear it off. Fig. 5-10. Remove the paper from inside the armor as far as you can. Then, with a sharp yank, tear it off *inside* the armor. Fig. 5-11. Once the paper has been removed, you must insert a fiber bushing around the armor and wires. Figs. 5-12 and 5-13. This protects the wires against damage. Their insulation could be punctured by the cut ends of the metal cable housing during installation.

Installing a connector on the armored cable can be done as shown in Fig. 5-14. The connector can then be placed inside a box as shown in Fig. 5-15. Drive the locknut of the connector down tightly. Make sure the teeth of the locknut bite into the metal of the box.

Armored cable must be supported every $4\frac{1}{2}$ feet by a method such as that shown in Fig. 5-16. In addition to this support it should be supported within 12 inches of each

5-17. *Placement of armored cable within a box and connecting it to a duplex receptacle.* *(Harvey Hubbell)*

GREEN HEX SCREW GROUNDS BOTH VOLTAGES

FOUR-WIRE BX

GROUNDING SLOT

250-VOLT TANDEM SLOTS

GROUNDING SLOT

125-VOLT PARALLEL SLOTS

WHITE

RED

BLACK

GROUNDING WIRE

COMBINATION RECEPTACLE

box. However, you can leave a free length not over 24 inches long at terminals where flexibility is needed. This may be the case with a motor that must be adjusted for belt tension. Keep in mind that the staples used must be rustproof. The NEC requires it.

Figure 5-17 shows how the BX cable is attached to the inside of a metal box for use in hooking up a combination duplex receptacle. This one will accomodate a 250-volt tandem plug and a 125-volt standard three-prong plug. Note the BX has four wires.

Rigid Metal Conduit

This method of wiring involves pipes that enclose the wires. The conduit is a pipe. It is aluminum or steel. This is not the same as water pipe. The inside of the pipe must be smooth so the wires can be pulled through without ruining the insulation. The steel pipe bends easily. The type of steel used allows for manual bending. The conduit may be black. If so, it is used indoors only in dry locations. The conduit also may be galvanized.

Some aluminum and some steel conduit have additional coatings of plastic to protect them. Aluminum conduit must have an additional plastic coating if it is to be used in contact with the earth or buried in concrete.

Conduit has a UL label. Fig. 5-18. All sizes are identical in dimensions with the corresponding size of water pipe, but the nominal size indicates the internal diameter with water pipe. However, the actual diameter is larger than water pipe. Conduit, for instance, of $\frac{1}{2}$-inch diameter will actually have a diameter closer to $\frac{5}{8}$ inch. The actual outside or external diameter will be about $\frac{7}{8}$ inch.

Use a special conduit cutter for reducing the conduit from its usual 10-foot length. A hacksaw will work. Do not use the pipe cutter. A pipe cutter will leave a sharp edge inside the conduit. This edge will scrape the insulation and cause shorts in the electrical wiring. A reamer should be used to make sure the cut surfaces are smooth.

Figure 5-19 shows how a piece of rigid conduit is used to support a box that has a duplex receptacle. Note the grounding arrangement in addition to the pipe itself.

Figure 5-20 shows the use of rigid conduit for a residential installation of the service entrance. Note the straps used to hold the rigid conduit securely.

White

These three lead
wires must extend
out of entrance
head at least 36 in.

Black

Red

Rigid Conduit

Conduit
strap

Conduit
connector

Meter
socket

Rigid or
thin-wall
conduit

Conduit
adapter

Entrance
ell

Conduit
connector

5-18. *Rigid metal conduit.*

Underwriters Laboratories Inc.
LISTED
RIGID CONDUIT
ISSUE SA-11

WHITE

BLACK

RECEPTACLE

GROUNDING
SLOTS

GREEN HEX
GROUNDING SCREW

GROUNDING WIRE
(GREEN OR GRAY)

5-19. *Placement of rigid conduit into a box to
hold a duplex receptacle. (Harvey Hubbell)*

5-20. *Using conduit for a service
entrance. (Sears)*

Service panel

Table 5-A. Maximum Number of Conductors in Trade Sizes of Conduit or Tubing*

Type letters	Conductor size AWG, MCM	1/2	3/4	1	1 1/4	1 1/2	2	2 1/2	3	3 1/2	4	4 1/2	5	6
TW, T, RUH,	14	9	15	25	44	60	99	142						
RUW,	12	7	12	19	35	47	78	111	171					
XHHW (14 thru 8)	10	5	9	15	26	36	60	85	131	176				
	8	2	4	7	12	17	28	40	62	84	108			
RHW and RHH	14	6	10	16	29	40	65	93	143	192				
(without outer	12	4	8	13	24	32	53	76	117	157				
covering),	10	4	6	11	19	26	43	61	95	127	163			
THW	8	1	3	5	10	13	22	32	49	66	85	106	133	
TW,	6	1	2	4	7	10	16	23	36	48	62	78	97	141
T,	4	1	1	3	5	7	12	17	27	36	47	58	73	106
THW,	3	1	1	2	4	6	10	15	23	31	40	50	63	91
RUH (6 thru 2),	2	1	1	2	4	5	9	13	20	27	34	43	54	78
RUW (6 thru 2),	1		1	1	3	4	6	9	14	19	25	31	39	57
FEPB (6 thru 2),	0		1	1	2	3	5	8	12	16	21	27	33	49
RHW and	00		1	1	1	3	5	7	10	14	18	23	29	41
RHH (without	000		1	1	1	2	4	6	9	12	15	19	24	35
outer covering)	0000			1	1	1	3	5	7	10	13	16	20	29
	250			1	1	1	2	4	6	8	10	13	16	23
	300			1	1	1	2	3	5	7	9	11	14	20
	350				1	1	1	3	4	6	8	10	12	18
	400				1	1	1	2	4	5	7	9	11	16
	500				1	1	1	1	3	4	6	7	9	14
	600					1	1	1	3	4	5	6	7	11
	700					1	1	1	2	3	4	5	7	10
	750					1	1	1	2	3	4	5	6	9

*Reprinted by permission from NFPA 70-1981, *National Electrical Code,* © 1980 National Fire Protection Association, Boston, Massachusetts

The number of conductors in a piece of conduit is regulated by the Code. The number of conductors permitted in a particular size of conduit or tubing is covered in Chapter 9 of the Code. Tables 5-A and 5-B. These tables cover conductors of the same size used for either new work or rewiring. Tables 5-C and 5-D cover combinations of conductors of different sizes when used for new or rewiring. In most cases the Code does not permit over 40% of the conduit area to be filled with wires.

A conduit run should be installed so that no injury comes to the wires being pulled through it. The basic rule for support of rigid conduit is within every 3 feet of a connection to an enclosure or fitting and then every 10 feet. If threaded couplings are used, then wider spacing for supports can be used.

Table 5-B. Maximum Number of Conductors in Trade Sizes of Conduit or Tubing

Type letters	Conductor size AWG, MCM	1/2	3/4	1	1 1/4	1 1/2	2	2 1/2	3	3 1/2	4	4 1/2	5	6
THWN	14	13	24	39	69	94	154							
	12	10	18	29	51	70	114	164						
	10	6	11	18	32	44	73	104	160					
	8	3	5	9	16	22	36	51	79	106	136			
THHN	6	1	4	6	11	15	26	37	57	76	98	125	154	
	4	1	2	4	7	9	16	22	35	47	60	75	94	137
FEP (14 thru 2)	3	1	1	3	6	8	13	19	29	39	51	64	80	116
FEPB (14 thru 8)	2	1	1	3	5	7	11	16	25	33	43	54	67	97
	1			1	3	5	8	12	18	25	32	40	50	72
XHHW (4 thru 500MCM)	0		1	1	3	4	7	10	15	21	27	33	42	61
	00		1	1	2	3	6	8	13	17	22	28	35	51
	000		1	1	1	3	5	7	11	14	18	23	29	42
	0000		1	1	1	2	4	6	9	12	15	19	24	35
	250			1	1	1	3	4	7	10	12	16	20	28
	300			1	1	1	3	4	6	8	11	13	17	24
	350			1	1	1	2	3	5	7	9	12	15	21
	400				1	1	1	3	5	6	8	10	13	19
	500				1	1	1	2	4	5	7	9	11	16
	600				1	1	1	1	3	4	5	7	9	13
	700					1	1	1	3	4	5	6	8	11
	750					1	1	1	2	3	4	6	7	11
XHHW	6	1	3	5	9	13	21	30	47	63	81	102	128	185
	600					1	1	1	3	4	5	7	9	13
	700					1	1	1	3	4	5	6	7	11
	750					1	1	1	2	3	4	6	7	10
RHW	14	3	6	10	18	25	41	58	90	121	155			
	12	3	5	9	15	21	35	50	77	103	132			
	10	2	4	7	13	18	29	41	64	86	110	138		
	8	1	2	4	7	9	16	22	35	47	60	75	94	137
RHH (with outer covering)	6	1	1	2	5	6	11	15	24	32	41	51	64	93
	4	1	1	1	3	5	8	12	18	24	31	39	50	72
	3	1	1	1	3	4	7	10	16	22	28	35	44	63
	2		1	1	3	4	6	9	14	19	24	31	38	56
	1		1	1	1	3	5	7	11	14	18	23	29	42
	0		1	1	1	2	4	6	9	12	16	20	25	37
	00			1	1	1	3	5	8	11	14	18	22	32
	000			1	1	1	3	4	7	9	12	15	19	28
	0000			1	1	1	2	4	6	8	10	13	16	24

(Continued on following page)

Table 5-B. Maximum Number of Conductors in Trade Sizes of Conduit or Tubing (Continued)

Type letters	Conductor size AWG, MCM	1/2	3/4	1	1 1/4	1 1/2	2	2 1/2	3	3 1/2	4	4 1/2	5	6
RHW	250				1	1	1	3	5	6	8	11	13	19
	300				1	1	1	3	4	5	7	9	11	17
	350				1	1	1	2	4	5	6	8	10	15
	400				1	1	1	1	3	4	6	7	9	14
RHH (with outer covering)	500				1	1	1	1	3	4	5	6	8	11
	600					1	1	1	2	3	4	5	6	9
	700					1	1	1	1	3	3	4	6	8
	750					1	1	1	3	3	3	4	5	8

*Reprinted by permission from NFPA 70-1981, *National Electrical Code*, © 1980 National Fire Protection Association, Boston, Massachusetts

Table 5-C. Dimensions and Percent Area of Conduit and of Tubing*

Tables 5-C and 5-D give the nominal size of conductors and conduit or tubing for use in computing size of conduit or tubing for various combinations of conductors. The dimensions represent average conditions only. Variations will be found in dimensions of conductors and conduits of different manufacture.

Trade size	Internal diameter (inches)	Total 100%	Not lead covered			Lead covered				
			2 Cond. 31%	Over 2 cond. 40%	1 Cond. 53%	1 Cond. 55%	2 Cond. 30%	3 Cond. 40%	4 Cond. 38%	Over 4 cond. 35%
1/2	.622	.30	.09	.12	.16	.17	.09	.12	.11	.11
3/4	.824	.53	.16	.21	.28	.29	.16	.21	.20	.19
1	1.049	.86	.27	.34	.46	.47	.26	.34	.33	.30
1 1/4	1.380	1.50	.47	.60	.80	.83	.45	.60	.57	.53
1 1/2	1.610	2.04	.63	.82	1.08	1.12	.61	.82	.78	.71
2	2.067	3.36	1.04	1.34	1.78	1.85	1.01	1.34	1.28	1.18
2 1/2	2.469	4.79	1.48	1.92	2.54	2.63	1.44	1.92	1.82	1.68
3	3.068	7.38	2.29	2.95	3.91	4.06	2.21	2.95	2.80	2.58
3 1/2	3.548	9.90	3.07	3.96	5.25	5.44	2.97	3.96	3.76	3.47
4	4.026	12.72	3.94	5.09	6.74	7.00	3.82	5.09	4.83	4.45
4 1/2	4.506	15.94	4.94	6.38	8.45	8.77	4.78	6.38	6.06	5.56
5	5.047	20.00	6.20	8.00	10.60	11.00	6.00	8.00	7.60	7.00
6	6.065	28.89	8.96	11.56	15.31	15.89	8.67	11.56	10.98	10.11

*Reprinted by permission from NFPA 70-1981, *National Electrical Code*, © 1980 National Fire Protection Association, Boston, Massachusetts

Table 5-D. Properties of Conductors**

Tables 5-C and 5-D give the nominal size of conductors and conduit or tubing for use in computing size of conduit or tubing for various combinations of conductors. The dimensions represent average conditions only. Variations will be found in dimensions of conductors and conduits of different manufacture.

Size AWG MCM	Area cir. mils	Concentric lay stranded conductors		Bare conductors		DC resistance ohms/M ft. at 25° C [77° F.]		
		No. wires	Diam. each wire inches	Diam. inches	*Area sq. inches	Copper		Alumi-num
						Bare cond.	Tin'd. cond.	
18	1620	Solid	.0403	.0403	.0013	6.51	6.79	10.7
16	2580	Solid	.0508	.0508	.0020	4.10	4.26	6.72
14	4110	Solid	.0641	.0641	.0032	2.57	2.68	4.22
12	6530	Solid	.0808	.0808	.0051	1.62	1.68	2.66
10	10380	Solid	.1019	.1019	.0081	1.018	1.06	1.67
8	16510	Solid	.1285	.1285	.0130	.6404	.659	1.05
6	26240	7	.0612	.184	.027	.410	.427	.674
4	41740	7	.0772	.232	.042	.259	.269	.424
3	52620	7	.0867	.260	.053	.205	.213	.336
2	66360	7	.0974	.292	.067	.162	.169	.266
1	83690	19	.0664	.332	.087	.129	.134	.211
0	105600	19	.0745	.372	.109	.102	.106	.168
00	133100	19	.0837	.418	.137	.0811	.0843	.133
000	167800	19	.0940	.470	.173	.0642	.0668	.105
0000	211600	19	.1055	.528	.219	.0509	.0525	.0836
250	250000	37	.0822	.575	.260	.0431	.0449	.0708
300	300000	37	.0900	.630	.312	.0360	.0374	.0590
350	350000	37	.0973	.681	.364	.0308	.0320	.0505
400	400000	37	.1040	.728	.416	.0270	.0278	.0442
500	500000	37	.1162	.813	.519	.0216	.0222	.0354
600	600000	61	.0992	.893	.626	.0180	.0187	.0295
700	700000	61	.1071	.964	.730	.0154	.0159	.0253
750	750000	61	.1109	.998	.782	.0144	.0148	.0236
800	800000	61	.1145	1.030	.833	.0135	·.0139	.0221
900	900000	61	.1215	1.090	.933	.0120	.0123	.0197
1000	1000000	61	.1280	1.150	1.039	.0108	.0111	.0177
1250	1250000	91	.1172	1.289	1.305	.00863	.00888	.0142
1500	1500000	91	.1284	1.410	1.561	.00719	.00740	.0118
1750	1750000	127	.1174	1.526	1.829	.00616	.00634	.0101
2000	2000000	127	.1255	1.630	2.087	.00539	.00555	.00885

*Area given is that of a circle having a diameter equal to the over-all diameter of a stranded conductor.
**Reprinted by permission from NFPA 70-1981, *National Electrical Code*, © 1980 National Fire Protection Association, Boston, Massachusetts

5-21. *Note the support strap used near the connector and the coupling. (Sears)*

Connector

Conduit

Strap

Coupling

EMT

5-22. *Electrical metallic tubing (EMT).*

Figure 5-21 shows how a strap is used to support a piece of conduit near the coupling and near the connector.

Nonmetallic Conduit

Plastic conduit is called nonmetallic. This plastic conduit is listed in sizes from $\frac{1}{2}$ inch to 6 inches. It is used primarily where there may be a corrosive atmosphere that would damage metal conduit. It can be used above ground for circuits operating with voltages up to 600 volts. It can be used as a general purpose raceway. Only PVC (polyvinyl chloride) is acceptable for use above ground in buildings.

It should be kept in mind that nonmetallic conduit is not permitted in ducts, plenums, and other air handling spaces.

Support rules for nonmetallic conduit are simple and direct. A clamp or other support must be placed within 3 feet of any box, cabinet, or other conduit termination. Spacing between supports varies with the temperature rating and the size of the conduit. See Section 347 of the Code for exact support spacing. It may vary from 2 to 8 feet.

All cuts should be trimmed inside and out and the surfaces made smooth. Couplings and enclosures must be of the type approved for the conduit. In plastic conduit the cement used to make connections should also be of the type approved by the Code or local ordinance.

Electrical Metallic Tubing (EMT)

This type of conduit is thinner in outside wall thickness than rigid conduit. The internal diameter in the smaller sizes is the same as in rigid conduit. In the larger sizes, the internal diameter is a little larger. The walls are thin and should not be threaded. All fittings are threadless. They hold to the material through pressure. Figure 5-22 shows a piece of EMT. Figure 5-23 shows a connector and a coupler for use with thinwall conduit.

5-23. *A connector and a coupling used with thinwall conduit (EMT).*

To cut thinwall conduit, use a hacksaw with 32 teeth to the inch. A metal removing power saw can also be used. The burrs and sharp edges should be removed with a reamer. A complete set of fittings and methods of using EMT will be seen in another chapter in this book.

Flexible Metal Conduit

EMT is restricted in its use to where it will not be exposed to physical abuse or corrosive agents. Flexible metal conduit is also designed for use where it will not be subject to physical damage. Fig. 5-24.

A typical use of flexible metallic tubing is a 4- to 6-foot length for a fixture whip in a ceiling. Fig. 5-25. The conduit may contain two No. 18 Type AF wires for a 6-ampere

5-24. *Flexible metal conduit. Wires are pulled through this conduit as with any other type.*

5-25. *Using flex tubing for a light fixture.*

Lighting supplied by branch circuit protected at not over 20A

Outlet box

Conduit or cable circuit

May be 3/8-in. size

BEND

Suspended ceiling

Recessed incandescent fixture

fixture load. Section 240-4 of the *Code* allows No. 16 and No. 18 fixture wire to be protected at 20 amperes. The flex tubing is also the equipment grounding conductor. It should be limited to dry locations if it does not have a nonmetallic coating. It cannot be used for direct earth burial or for embedding in poured concrete or aggregate. It is also limited to lengths of not more than 6 feet. This means it is ruled out from being used as a general-purpose raceway. It has not been judged as being adequate for grounding purposes for lengths over 6 feet. The ⅜-inch flexible tubing usually can handle two No. 14 type AF wires and no grounding conductor. It is primarily used for attaching light fixtures. Straps or supports should be placed within 12 inches of each end and 4½ feet in between. Unclamped lengths of *flex* is permitted by an exception to the *Code.*

For concealed work, the run of flex from one outlet to another or to a fitting must not have more than the equivalent of four quarter-bends. This includes those bends located at the outlet or fitting. Angle connectors cannot be used for concealed flex installations. Straight connectors, however, can be used.

QUESTIONS

1. What is the NEC?
2. What does the symbol UL mean?
3. What does the symbol CSA mean?
4. What does the Code do for electrical safety?
5. What is the difference between rigid conduit and armored cable?
6. What controls the number of conductors permitted to be in a conduit?
7. Where is nonmetallic (plastic) conduit used?
8. What does EMT mean?
9. What is thinwall conduit?
10. How does thinwall conduit differ from rigid conduit?
11. What is meant by flexible metal conduit?
12. Where is flexible metal conduit used?

CHAPTER 6

Planning

In new construction, the residential electrician has to follow the plans given him or her. Any changes must be cleared with the subcontractor, the contractor and the owner. In remodeling, the electrician is working, in most cases, directly with the owner. Decisions can be made at the time a problem arises. However, there are some very basic things to be considered in making a plan a reality. There are code restrictions. Codes are local in nature and require various types of hook-ups, wire sizes, and boxes. The method of installation is very important for the safety of the resident.

The National Electrical Code (NEC) is used as a basic standard when there is no other code to be followed. In this chapter we will deal with the requirements for a good living environment. At the same time, we will point out some of the code restrictions that should be used as a basis for wiring the house.

Planning is important so that the proper amount of materials and equipment are handy when needed. Once the job is finished, the work should pass inspection by a person other than the one who installed it. In most communities the local electrical inspector is called to check the wiring before it is enclosed behind drywall and after it is finished and operating as a system.

The correct wire size, from the transformer on the pole to the outlet on the wall, helps prevent such occurrences as blinking lights when the refrigerator turns on. Wire size is determined by the amount of current required by power consuming devices. These devices present a load to the power source. The load is usually shown in watts on appliance nameplates. Power, which is measured in watts, is determined by multiplying the voltages times the current. If an overload exists, or if the wires are too small, a voltage drop caused by the small wires reduces power available inside the house.

The larger the physical size of the wire, the more current it can handle without line losses caused by an increase in the load. Wire size is stated in numbers. The larger the number, the smaller the wire.

FIGURING CIRCUIT LOAD

The load presented by a circuit may be figured by simply adding the wattage ratings of the devices to be attached to the line or plugged into the circuit. When planning a new house or rewiring an old one, it is best to figure the amount of wattage needed and then design the circuits to accommodate the load.

There are a number of reasons why you need adequate wiring in a house—new or old. If fuses blow, or a circuit breaker trips, probable overloads exist. If the lights dim, blink, or flicker whenever an additional device is turned on, then the circuits are overloaded. A television picture will shrink if it is on a line with an overload. If it is color television, the colors may change when something else turns on in the house. If an electric iron or electric range takes longer than usual to heat up, then chances are the circuits are overloaded. Air conditioners place quite a load on available circuitry. If the air conditioner is more than 5000 Btu's, it should be on a separate circuit.

Rewiring may be necessary to eliminate some of these problems. However, at the time a new house is being planned, it is best to make sure future electrical needs are taken into consideration. During the past twenty years, the use of electricity in the home has more than tripled. Probably, the future will hold more demands for electricity. Adequate planning is a must if blinking lights, shrinking TV pictures, and blown fuses are to be prevented.

General Requirements for Residential Wiring

A fundamental prerequisite for any adequate wiring installation is conformity to the safety regulations applicable to the dwelling. The approved American standard for electrical safety is the National Electrical Code. Usually, local safety regulations in municipal ordinances or state laws are based on the National Electrical Code.

Inspection service, to determine conformance with safety regulations, is available in most communities. Where available, a certificate of inspection should be obtained. In the absence of inspection service, a certificate should be obtained from the electrical contractor who installed the wiring. The certificate should say the wiring conforms with the applicable safety regulations.

LOCATING LIGHTING OUTLETS. Proper illumination is an essential element of modern living. The amount and type of illumination required should be adapted to the various seeing tasks in the home. There must also be planned lighting for recreation and entertainment. In many instances, best results are achieved by a blend of lighting from both fixed and portable light sources. Good home lighting, therefore, requires thoughtful planning and careful selection of fixtures, portable lamps, and other lighting equipment.

Where lighting outlets are mentioned in plans, their types and locations should conform with the lighting fixtures or equipment to be used. Unless a specified location is stated, lighting outlets may be located anywhere within the area under consideration.

LOCATING CONVENIENCE OUTLETS. Convenience outlets should be located near the ends of wall space, rather than near the center. This reduces the likelihood of their being concealed behind large pieces of furniture. Unless otherwise specified, outlets should be located approximately 12 inches above the floorline.

LOCATING WALL SWITCHES. Wall switches should normally be located at the latch side of the doors, or at the traffic side of arches, and within the room or area to which the control is applicable. Some exceptions to this practice are: (1) control switches for exterior lights from indoors, (2) control of stairway lights from adjoining areas where stairs are closed off by doors at head or foot, and (3) the control of lights from the access space adjoining infrequently used areas, such as storage rooms. Wall switches are normally mounted approximately 48 inches above the floor.

CONTROLLING FROM MORE THAN ONE SWITCH. All spaces for which wall switch controls are required, and which have more than one entrance, should be equipped with a multiple-switch control at each principal entrance. If this requirement would result in the placing of switches controlling the same light within 10 feet of each other, one of the switch locations may be eliminated.

Some rooms are lighted from more than one source, as when both general and supplementary illumination is provided from fixed sources. In this case, multiple switching is required for only one set of controls, usually to the general illumination circuit.

Principal entrances are those commonly used for entry to, or exit from, a room when going from a lighted to an unlighted condition, or the reverse. For instance, a door from a living room to a porch is a principal entrance to the porch. However, it would not necessarily be considered a principal entrance to the living room, unless the front entrance to the house is through a porch.

Number of Local Lighting or Convenience Outlets Necessary

If the standards require a certain number of outlets based on a linear or square-foot measure, the number of outlets shall be determined by dividing the total linear or square footage by the required distance or area. The number so determined should be increased by one if a major fraction remains.

Example:

> Required: One outlet for each 150 square feet.
> Total square feet of area: 390
> 390 divided by 150 equals 2.6
> Outlets required: 3

DUAL-PURPOSE ROOMS. A room may be intended to serve more than one function, such as a combination living-dining room or a kitchen-laundry. In this case, convenience and special-purpose outlet provisions of these standards are separately applicable to those respective areas. Lighting outlet provisions may be combined in any manner that will assure general, overall illumination, as well as local illumination of work surfaces. In locating wall switches, a dual-purpose area is considered a single room.

DUAL FUNCTIONS OF OUTLETS. In any instance where an outlet is located to satisfy two different provisions of the standard of adequacy, only one outlet need be installed at that location. In such instances, particular attention should be paid to required wall switch controls, as additional switching may be necessary.

For example, a lighting outlet in an upstairs hall may be located at the head of the stairway. This satisfies both a hall-lighting outlet and a stairway-lighting outlet provision of Code standards with a single outlet. Stairway-lighting provisions will necessitate multiple-switch control of this outlet at both the head and the foot of the stairway. In addition, if the hallway is long, the multiple-switch control rule, previously stated, may require a third point of control in the hall.

INTERPRETING REQUIREMENTS

The following standards are necessarily general in nature. They apply to most situations encountered in normal house construction and are to be considered as *minimum* standards of adequacy for such construction. Situations will arise, because of unusual design or unusual construction methods, when it will be impossible to satisfy a particular provision of the standards. The following rules may serve as a guide for meeting such situations:

6-1. *Wiring requirements of a typical home.*

• Wiring installation should be fitted to the structure. If compliance with a particular provision would require alteration of doors, windows, or structural members, alternate wiring provisions should be made. For example, some types of building construction may make it desirable to alter the recommended outlet or switch heights presented here, or to resort to surface-type wiring construction.

• Each provision of the adequacy standard is intended to provide for one or more specific usages of electricity. If such usage is appropriate to the particular home under study, and cannot be provided in accordance with the standard, an alternate pro-

6-2. *Outlet requirements for a basement of a typical home.*

SKETCH OF THREE-LEVEL HOUSE

6-3. *Electrical wiring diagram for a floor of a home.*

vision should be made. If a construction peculiarity eliminates the need for a particular electrical facility, no alternate provision is needed.

● If certain functions are omitted from the plans, such as a workshop or a laundry, electrical wiring serving these functions naturally would be omitted.

● If certain facilities are indicated as future additions to the plans, for example, a basement recreation room, the initial wiring should be so arranged that none of it need be replaced, or moved, when the ultimate plan is realized. Final wiring for the future addition may be left to a later date, if desired.

● Standards given here may be applied to multifamily dwelling units. At the present time there are no nationally recognized standards for common-use spaces in such buildings. This makes good judgment on the part of the planner of electrical wiring a necessity. Particular attention should be paid to limiting voltage drop in feeders to individual apartments in order to assure proper operation of appliances and space-heating equipment.

TYPICAL OUTLET REQUIREMENTS FOR A HOME

Figures 6-1 through 6-5 show a method of drawing in outlets and switches on the floor plan of a house. The symbols used are identified in Table 6-A. The use of these

6-4. *The second level of the home and its electrical devices.*

symbols makes it easier to sketch in the requirements before the building is started. The homeowner can then check these requirements. In the case of remodeling, the symbols can be used to show where new wiring and new electrical devices will be installed. This preplanning makes it much easier for the electrician to figure out the amount of wire and the devices needed for a job. Such preplanning saves time and energy.

Other considerations to be made in the wiring plan of a house include doorbells, chimes, communications centers (intercoms), household fire-warning systems, and televisions with special lead-ins and rotor wires. These items are discussed in other chapters of this book.

6-5. *Convenience outlets for a basement of a house.*

Table 6-A. Graphical Electrical Symbol for Residential Wiring Plans

General Outlets

Lighting Outlet

Ceiling Lighting Outlet for recessed fixture (Outline shows shape of fixture.)

Continuous Wireway for Fluorescent Lighting on ceiling, in coves, cornices, etc. (Extend rectangle to show length of installation.)

Lighting Outlet with Lamp Holder

Lighting Outlet with Lamp Holder and Pull Switch

Fan Outlet

Junction Box

Drop-Cord Equipped Outlet

Clock Outlet

To indicate wall installation of above outlets, place circle near wall and connect with line as shown for clock outlet.

Convenience Outlets

single outlet

Duplex Convenience Outlet

Triplex Convenience Outlet (Substitute other numbers for other variations in number of plug positions.)

Duplex Convenience Outlet — Split Wired

Weatherproof Convenience Outlet

Multi-Outlet Assembly (Extend arrows to limits of installation. Use appropriate symbol to indicate type of outlet. Also indicate spacing of outlets as X inches.)

Combination Switch and Convenience Outlet

Combination Radio and Convenience Outlet

Floor Outlet

Range Outlet

Special-Purpose Outlet. Use subscript letters to indicate function. DW-Dishwasher, CD-Clothes Dryer, etc.

Space Heating and Air Conditioning Outlets
Use Special-Purpose Outlet and Subscripts

230-v Space Heating Outlet

Space Heating Furnace

Space Heating Boiler

Heat Pump

Central Air Conditioning

230-v Air Conditioning Outlet

Switch Outlets

S — Single-Pole Switch

S_3 — Three-Way Switch

S_4 — Four-Way Switch

S_D — Automatic Door Switch

S_P — Switch and Pilot Light

S_{WP} — Weatherproof Switch

S_2 — Double-Pole Switch

Low-Voltage and Remote-Control Switching Systems

S — Switch for Low-Voltage Relay Systems

MS — Master Switch for Low-Voltage Relay Systems

O_R — Relay—Equipped Lighting Outlet

Low-Voltage Relay System Wiring

Auxiliary Systems

Push Button

Buzzer

Bell

Combination Bell-Buzzer

CH — Chime

Annunciator

D — Electric Door Opener

M — Maid's Signal Plug

Interconnection Box

T — Bell-Ringing Transformer

Outside Telephone

Interconnecting Telephone

R — Radio Outlet

TV — Television Outlet

Miscellaneous

Service Panel

Distribution Panel

Switch Leg Indication. Connects outlets with control points.

Special Outlets. Any standard symbol given above may be used with the addition of subscript letters to designate some special variation of standard equipment for a particular architectural plan. When so used, the variation should be explained in the Key of Symbols and, if necessary, in the specifications.

129

6-6. A weatherproof outlet, preferably near the front entrance, should be located at least 457 mm [18 in] above grade and be GFI protected.

LIVING ROOM

TO DINING ROOM

6-7. Some means of general illumination is essential. It may be provided by ceiling or wall fixtures, by lighting in covers, valances, or cornices, or by portable lamps. Provide lighting outlets, wall-switch controlled, in locations appropriate to the lighting method selected.

Planning Requirements for Individual Areas

EXTERIOR ENTRANCE WIRING. The following planning requirements should be observed:

• *Lighting provisions.* It is recommended that lighting outlets, controlled from wall switches, be installed at all entrances. Fig. 6-6. The principal lighting requirements for entrances are the illumination of steps leading to the entrance, and of faces of people at the door.

• *Convenience outlets.* Outlets in addition to those at the door are often desirable for post lights to illuminate terraced or broken flights of steps, or long approach walks. These outlets should be wall-switch controlled from inside the house, near the entrance.

It is recommended that an exterior outlet near the front entrance be controlled by a wall switch inside the entrance. This provides for convenient operation of outdoor lighting, lawn mowers, and hedge trimmers.

• *Code requirements.* The NEC requires a ground fault interrupter (GFI) if the outlet near the front entrance is outside. The rule requires GFI protection if the receptacle is where there is direct grade level access to the dwelling unit and to the receptacles. Thus, a GFI would not be required on a porch or raised portion of a one-family residence. Balconies of apartments and condominiums would not require a GFI installed in the outside receptacle.

Receptacle required if any
wall width is 2 ft or more

6-8. *Location of wall outlets
in a living room.*

Living
room

Walls with sliding
glass doors or
floor-to-ceiling
windows must be
counted as wall
space requiring
receptacles

Wall
receptacles

Floor receptacles must be
close to wall to count as
required spaced receptacles

LIVING ROOM. The following planning requirements should be observed:

• *Lighting provisions.* Provisions for general illumination of living rooms also apply to sun rooms, enclosed porches, family rooms, television rooms, libraries, dens, and similar living areas. Fig. 6-7. Installation of outlets for decorative lighting accents is recommended, such as picture illumination and bookcase lighting.

One of the requirements for a living room is light. The placement of floor lamps and table lamps is important for reading and enjoying the space. Therefore, it is necessary to provide outlets that will be easily accessible to the lamps.

Receptacles are required on any wall with a width of more than two feet. For example, receptacles are not required on a wall with a width of 23 inches. Walls with sliding glass doors or floor-to-ceiling windows must be counted as a wall space that requires receptacles. Here, floor receptacles may need to be installed. Fig. 6-8. From any point along the wall, at the floor line, a receptacle must be not more than 6 feet away.

Figure 6-9 shows the location of a receptacle that will allow a lamp to be plugged in with no more than 6 feet from either side of the receptacle.

6-9. *Location of a receptacle that will permit plugging in of a lamp or appliance.*

Extension cords should not be used where people can walk over them.

Electric heaters may be run along the wall and take up floor space when permanently mounted.

Making the receptacle part of the electric heater can ensure that the extension cord or lamp cord does not fall into the path of hot air.

Fixed room dividers, especially in the living room, must be counted in considering the spacing of receptacles. Figure 6-10 shows how this is done. Since the two sides of the room divider provide room space, a table lamp may be located where it would not be possible to plug it in without an extension cord. It is possible to place a receptacle on each side of the divider. It is also possible to place a receptacle inside the divider itself.

For switch controlled outlets, it is recommended that split-receptacle outlets be used. Their use does not limit the location of radios, television sets, clocks and similar devices. Figure 6-11 shows that one convenience outlet should be installed flush in the mantel shelf if a fireplace is included in the room.

• *Special-purpose outlets.* It is recommended that one outlet for a room air conditioner and outlets for space heating be installed if central systems are not planned.

DINING AREAS. The following planning requirements should be observed:

• *Lighting provisions.* Each dining room, or dining area combined with another room, or breakfast nook shall have at least one lighting outlet, wall-switch controlled. Fig. 6-12. When a dining or breakfast table is to be placed at the table location, the switch should be placed just above table height. Built-in counter space should have an outlet provided above counter height for portable appliances.

• *Convenience outlets.* These should be of the split-receptacle type, for connection to small appliances.

6-10. *Fixed wall dividers are counted as wall space when figuring the number of outlets needed.*

6-11. *Location of convenience outlets in a living room.*

6-12. Lighting outlets for dining areas are normally located over the probable location of the dining or breakfast table to provide direct illumination.

6-13. Good general illumination is particularly important in a bedroom. This can be provided from a ceiling fixture or from lighting valances, coves, or cornices. Provide the proper wall switch controls.

6-14. Start with an imaginary centerline through each probable bed location. Install outlets on each side of the centerline and within 6 feet of it. No point along the floorline in any other usable wall space should be more than 6 feet from an outlet on that wall. Add outlets as needed to achieve this.

● *Special outlets.* It is recommended that outlets for space heating and one outlet for a room air conditioner be installed whenever central systems are not planned.

BEDROOMS. The following planning requirements should be observed:

● *Lighting provisions.* Light fixtures over full-length mirrors, or a light source at the ceiling located in the bedroom and directly in front of the clothes closets, may serve as general illumination. Fig. 6-13.

Master-switch control in the master bedroom, as well as at other strategic points in the home, is suggested for selected interior and exterior lights.

● *Convenience outlets.* It is recommended that convenience outlets be placed only 3 to 4 feet from the centerline of the probable bed locations. The popularity of bedside radios and clocks, lamps, and electric bed covers makes increased plug-in positions at bed locations essential. Triplex or quadruplex convenience outlets are therefore recommended at these locations. Fig. 6-14.

6-15. *Wiring diagrams for bedroom with twin beds.*

Furthermore, at one of the switch locations, there should be a receptacle outlet for a vacuum cleaner, floor polisher, or other portable appliances. Figs. 6-15 and 6-16.

• *Special-purpose outlets.* It is recommended that outlets for space heating and one outlet for a room air conditioner be installed in each bedroom if central systems are not provided.

BATHROOMS AND LAVATORIES. The following planning requirements should be observed:

• *Lighting provisions.* A ceiling outlet located in line with the front edge of the washbasin will provide lighting for the mirror, general room lighting, and safety lighting for a combination shower and tub. Figs. 6-17 and 6-18.

When more than one mirror location is planned, equal consideration should of course be given to the lighting in each case. A switch-controlled night-light may be installed where desired.

If an enclosed shower stall is planned, an outlet for a vapor-proof luminaire is sometimes installed, controlled by a wall switch outside the stall.

• *Convenience outlets.* It is suggested that an outlet be installed at each separate mirror or vanity space. An outlet should also be installed at any space that might accommodate an electric towel dryer, electric razors, or electric toothbrushes. A receptacle that is part of a bathroom lighting fixture should not be considered as satisfying this requirement for small appliances, unless it is rated at 15 amperes and wired with at least 15-ampere rated wires.

• *Code requirements.* Any receptacle that is an integral part of an appliance, cabinet, or lighting fixture may not be used to satisfy the specific receptacle requirements for bathrooms. Every bathroom must have at least one wall-switch controlled lighting outlet. This does not mean a pullchain or a canopy will serve as the requirement.

Take a close look at Fig. 6-19 to see what is and what is not considered a bathroom according to Code requirements. A receptacle is not usually required in the area with the tub and toilet. However, if one is installed in these areas it must be protected by a GFI.

6-16. The same bedroom with wiring adaption to double bed arrangement.

6-17. Illumination of both sides of the face when at a mirror is important. Keep in mind that a single concentrated light, either on the ceiling or side wall, is usually not acceptable. All lighting outlets should be wall-switch controlled.

6-18. One outlet near the mirror, 3 to 5 feet above the floor is required. See Code requirements about GFI and outlets.

Toilet and sink

BATHROOM

Sink plus tub and/or shower

BATHROOM

6-19. What is and what is not a bathroom as far as the Code is concerned.

Toilet only

NO BATHROOM

Tub and toilet

NO BATHROOM

Sink only

NO BATHROOM

Figure 6-20 shows a typical bedroom suite in a one-family house or apartment. Note the location of the bathroom off the bedroom. Even with the basin in the vanity outside the tub or toilet area it does require a receptacle at the basin A receptacle is not required in the room where the tub is located. Figure 6-21 indicates possible receptacle locations.

KITCHEN. The following planning requirements should be observed:

● *Lighting provisions.* Provide outlets for general illumination and for lighting at the sink. These lighting outlets should be wall-switch controlled. Lighting design should provide for illumination of work areas, sink, range, counters, and tables. Under-cabi-

6-20. *A typical bedroom suite with a bathroom off the bedroom. Note the location of the receptacle.*

6-21. *These are possible locations of the outlet for the bathroom. The location will vary, depending on the wall space available. A receptacle in the medicine cabinet or in the bathroom light does not count as the required receptacle.*

6-22. *Wiring diagram for a kitchen.*

6-23. *The following outlets are usually necessary: one for a free-standing range, or for a built-in range surface unit, and one for each built-in range oven (if each oven is a separate unit); one each for a ventilating fan, dishwasher, and food waste disposer (if necessary plumbing facilities are installed); and one for an electric clock.*

net lighting fixtures, within easy reach, may have local-switch control. In some kitchens, consideration should also be given to outlets to provide inside lighting of cabinets.

One outlet for the refrigerator is needed; and one outlet for each four linear feet of work-surface frontage, with at least one outlet to serve each work surface. If a planning desk is to be installed, one outlet is needed for this area. (Work-surface outlets to be located about 44 inches above the floorline.) Table space should have one outlet, preferably just above the table level. Fig. 6-22.

● *Convenience outlets.* An outlet is recommended at any wall space that may be used for an iron or for an electric toaster. Convenience outlets in the kitchen should be of the split-receptacle type, for connection to small appliances.

● *Special-purpose outlets.* One type of appliance which is helpful in the kitchen is a clock. It should be in a location easily visible from all parts of the kitchen. It requires a recessed receptacle, with a clock hanger. Fig. 6-23. An outlet for a food freezer, either in the kitchen or in some other convenient location, is often necessary.

● *Code requirements.* Receptacles are required at each counter space that is wider than 12 inches. If it is located inside the dining room, the counter must also have a receptacle if more than 12 inches of counter space is usable. Fig. 6-24.

6-24. *Required spacing for receptacles in a kitchen.*

Some receptacles are rendered inaccessible by the installation of appliances that are either fastened in place or positioned in a space that is dedicated or assigned permanently to an appliance. These receptacles are not counted as required outlets. The refrigerator would be an example of an appliance dedicated to a space.

A receptacle located behind an appliance and not accessible does not count as one of the required countertop receptacles. Countertop receptacles must be accessible.

LAUNDRY AND LAUNDRY AREAS. The following planning requirements should be observed:

• *Lighting provisions.* It is recommended that all laundry outlets be wall-switch controlled. Fig. 6-25.

6-25. *For a complete laundry, lighting outlets are installed over the work areas such as laundry tubs, sorting table, washing, ironing, and drying centers. For laundry trays in an unfinished basement: one ceiling outlet, centered over the trays. One convenience outlet, or more if desired, shall be installed.*

• *Convenience outlets.* Convenience outlets in the laundry area should be connected to a 20-ampere branch circuit that serves no other areas.

• *Special-purpose outlets.* Sometimes, plans require the installation of outlets for a ventilating fan and a clock.

• *Code requirements.* At least one receptacle, it may be a single, duplex or triplex, must be installed for the laundry of a dwelling. Such a receptacle and any other receptacles for special appliances must be placed within 6 feet of the intended location of the appliance. The Code requires a receptacle outlet in a basement in addition to any receptacle outlet that may be provided as the required receptacle to serve a laundry area in the basement. A separate receptacle must be provided for the appliance (usually a washing machine). One extra receptacle must be provided for the general use of anyone in the basement.

CLOSETS. The following planning requirements should be observed:

• *Lighting provisions.* One lighting outlet for each closet is suggested, except where shelving or other conditions make installation of lights within a closet ineffective. In this case, the lighting should be located in the adjoining space to provide light within the closet. Installation of wall switches near a closet door, or door-type switches, is recommended.

The light fixture must be so located that the bulb or fixture will not be within 12 inches of any shelf in the closet. Nothing piled up on the shelf should be within 12 inches of the lighting fixture. This is to prevent accidental fires from overexposure to the light source.

HALLS. The following planning requirements should be observed:

• *Lighting provisions.* Lighting outlets, wall-switch controlled, should be installed for proper illumination of the entire hall. Particular attention should be paid to irregularly shaped areas. These provisions apply to passage halls, reception halls, vestibules, entries, foyers, and similar areas. It is sometimes desirable to install a switch-controlled night-light in a hall with access to bedrooms.

• *Convenience outlets.* One convenience outlet for each 15 feet of hallway, measured along centerline is recommended. Each hall more than 25 square feet in floor area should have at least one outlet. In reception halls and foyers, convenience outlets shall be placed so that no point along the floorline in any usable wall space is more than 10 feet from an outlet in that space.

It is further recommended that at one of the switch outlets in the hallway, a convenience receptacle be provided for connection of a vacuum cleaner, or other floor-cleaning device.

• *Code requirements.* In hallways, stairways, and at outdoor entrances, remote, central, or automatic control of lighting shall be permitted according to the Code. This is an alternative to the wall-switch controlled light.

STAIRWAYS. The following planning requirement should be observed:

• *Lighting provisions.* Wall or ceiling outlets should provide adequate illumination of each stair flight. Outlets shall have multiple-switch control at the head and foot of the stairway, so arranged that full illumination may be turned on from either floor, but that lights in halls furnishing access to bedrooms may be extinguished without interfering with ground-floor usage.

These provisions are intended to apply to a stairway that connects finished rooms at either end.

Whenever possible, switches should be grouped together. They should never be located so close to steps that a fall might result while reaching for a switch.

At intermediate landings of a large stairway (depending on planned usage), an outlet is often recommended for a decorative lamp, night-light, or cleaning equipment.

RECREATION, TV, OR FAMILY ROOMS. The following planning requirements should be observed:

• *Lighting provisions.* Selection of a lighting method for these rooms should take into account the types of activities for which the room is planned. Fig. 6-26. General illumination is essential in a family, recreation, play, or TV room. It may be provided by ceiling or wall fixtures, or by lighting in coves, valances, or cornices. Provide lighting

FAMILY, RECREATION, PLAY, OR TV ROOM

6-26. *Lighting methods for recreation or family rooms should be selected according to the activities for which the room is planned.*

outlets, wall-switch controlled, in appropriate locations. Convenience outlets should be placed so that no point along the floorline in any usable wall space is more than 6 feet from an outlet.

• *Convenience outlets.* Convenience outlets in family rooms should be the split-receptacle type for connection to small appliance circuits.

If any of these rooms contains a fireplace, one convenience outlet should be installed flush in the mantel shelf. Outlets for the use of a clock, radio, television, ventilating fan, motion picture projector, and the like, should be located in relation to their intended use.

• *Special-purpose outlets.* Outlets for space heating and one outlet for a room air conditioner should be installed if central systems are not planned.

UTILITY ROOMS OR SPACE. The following planning requirements should be observed:

• *Lighting provisions.* Lighting outlets in a utility room (or space) should be placed to illuminate the furnace area, and work bench, if planned. At least one lighting outlet is to be wall-switch controlled. Fig. 6-27. One convenience outlet, preferably near the furnace, or near any planned workshop, should be installed. Special-purpose outlets, such as one each for the water heater, boiler, furnace, or other equipment, should be located convenient to the appliance or equipment.

• *Convenience outlets.* The washing machine outlet should be connected to a 20-ampere, laundry-area branch circuit.

BASEMENTS. The following planning requirements should be observed:

• *Lighting provisions.* In basements with finished rooms, with garage space, or with other direct access to outdoors, the stairway-lighting provisions stated previously apply. For basements that will be infrequently visited, a pilot light should be installed in conjunction with the switch at the head of the stairs.

• *Convenience outlets.* Basement lighting outlets should illuminate designated work areas or equipment locations, such as furnace, pump, and work bench. Fig. 6-28. Additional outlets should be installed near the foot of the stairway, in each closed space, and in open spaces so that each 150 square feet of open space is adequately served by a light in that area.

In unfinished basements, the light at the foot of the stairs should be wall-switched near the head of the stairs. Other lights may be chain-pulled for control. Convenience outlets in the basement should number at least two. For a work bench, one outlet should be placed at this location. In addition, one outlet for electrical equipment used in connection with furnace operation should be installed.

• *Special-purpose outlet.* If a food freezer is located here, an outlet for it must be installed. You may want to use the basement as a laundry or utility room. If so, check these headings for suggestions.

• *Code requirements.* At least one lighting outlet must be installed in every attic, underfloor space, utility room, and basement if it is used for storage or if it contains equipment requiring servicing. The lighting fixture does not necessarily have to be wall switched. A pull chain fixture is acceptable.

ACCESSIBLE ATTIC. The following planning requirements should be observed:

• *Lighting provisions.* One outlet for general illumination and a wall-switch controlled from the foot of the stairs is the minimum. When no permanent stairs are installed, this lighting outlet may be pull-chain controlled, if located over the access

6-27. *A wall-switched outlet is usually necessary to control the light in a utility room upon entering, because often this room has no windows.*

6-28. *Basement convenience outlets are useful near the furnace, at the play area, for basement laundries, darkrooms, hobby areas, and for appliances such as dehumidifiers and portable space heaters.*

door. Where an unfinished attic is planned for later development into rooms, the attic-lighting outlet shall be switch controlled at the top and bottom of the stairs. One outlet for each enclosed space is recommended.

These provisions apply to unfinished attics. For attics with finished rooms or spaces, see the appropriate room classification for requirements.

• *Convenience outlets.* One outlet for general use is the minimum. If an open stairway leads to future attic rooms, provide a junction box with direct connection to the distribution panel, for future extension to convenience outlets and lights when the rooms are finished.

A convenience outlet in the attic is desirable for providing additional light in dark corners, and also for the use of a vacuum cleaner and cleaning accessories.

• *Special-purpose outlet.* The installation of an outlet, switch controlled from a desirable point in the house, is recommended for use of a summer cooling fan.

PORCHES. The following planning requirement should be observed:

● *Convenience outlets.* Each porch, breezeway, or other similar roofed area of more than 75 square feet in floor area should have a lighting outlet, wall-switch controlled. Large or irregularly shaped areas may require two or more lighting outlets. One convenience outlet (weatherproof, if exposed to moisture) for each 15 feet of wall that borders a porch or breezeway is recommended. Fig. 6-29. Multiple-switch control should be installed at entrances if the porch is used as a passage between house and garage.

The split-receptacle convenience outlet is intended to be connected to a 3-wire appliance branch circuit. If this area is considered an outdoor dining area, all outside outlets must have a GFI (ground-fault interrupter). At least one outside outlet is required for any newly constructed house.

6-29. One outlet, weatherproof if exposed to moisture, for each 15 feet of wall bordering the porch or breezeway, is recommended. If the outlet is accessible by someone standing on the earth, then it must also be GFI protected. One or more such outlets should be controlled by a wall switch inside the door.

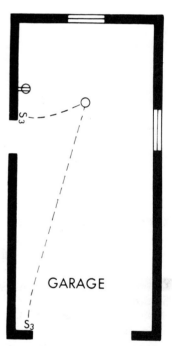

6-30. The garage or carport should contain at least one ceiling outlet, wall-switched, for a one- or two-car garage or storage area. If the garage has no covered access from the house, provide one exterior outlet, multiple-switch controlled, from both garage and residence. One plug-in outlet for a one- or two-car storage area is necessary. The outlet should be GFI protected.

TERRACES AND PATIOS. The following planning requirements should be observed:

• *Lighting provisions.* One or more outlets on the building wall, or other convenient location in the area to provide fixed illumination, are recommended. Wall-switch control shall be provided inside the house.

• *Convenience outlets.* One weatherproof outlet located at least 18 inches above the grade line is needed for each 15 linear feet of house wall bordering terrace or patio. It is recommended that one or more of these outlets be wall-switch controlled from inside the house.

GARAGES OR CARPORTS. If a garage is to be used for additional purposes, such as a work area, storage closets, laundry, attached porch, or the like, the rules apply which are appropriate to those uses. Fig. 6-30.

An exterior outlet, wall-switch controlled, is recommended for all garages. Additional exterior outlets are often desirable, even if no specific additional use is planned for the garage Long driveways warrant additional illumination, such as a post light, wall-switch controlled from inside the house.

If a food freezer or refrigerator, work bench, automatic door opener, or car heater is planned for the garage, outlets for these uses should be provided.

• *Code requirements.* A vehicle door in an attached garage is not considered as an outdoor entrance. This means the Code does not require a light outlet at the garage door. It is provided with light by the car's headlights.

If it is an attached garage there must be at least one receptacle. It should have GFI protection. GFI protection is required for all 120-volt, single-phase, 15- and 20-ampere receptacles installed in a garage. This is because homeowners use outdoor appliances plugged into the garage outlet. This gives protection for lawnmowers, snowblowers, hedge trimmers and hand-held electric drills. A carport attached to the house should have a GFI-protected outlet.

EXTERIOR GROUNDS. In addition to terrace and patio lighting, outlets under the eaves on the exterior of a house and garage are sometimes recommended for protective or decorative lighting. Switch control should be from within the house. Multiple and master-switch control is desirable also.

Weatherproof floodlights and spotlights are available to show the beauty of a garden by night. These are designed for use with either 115-volt lamps, or lamps at reduced voltage supplied through safety transformers. Fixtures may be permanently wired, and switched from inside the house, or they may be served from outdoor weatherproof convenience outlets. Lighting for sports (such as tennis, volleyball, or badminton) deserves consideration where the grounds are suitable for this purpose.

Underwater and area lighting for a swimming pool, and power requirements for a pool circulating pump and filter require careful design and installation by specialists in this field.

In general, it is recommended that all outdoor lighting outlets be switch controlled from within the house. In addition, certain ones, such as post lights and protective lighting, lend themselves also to photoelectric-cell or time switch control.

In climates where build-up of snow and ice on roofs is a problem, a "snow-melting" (heating) cable may be installed at the lower edge of the roof and in the gutters and downspouts. This provides excellent protection at low cost against water leakage into ceilings and outside walls of living quarters.

Motor-operated valves for buried lawn sprinkler systems are available. These can be arranged for manual or automatic operation. However, the water-operated ones are quieter.

It is best to check the architect's specifications for outlets with the proper person to make sure the proper number and size are installed. The best way to check is to ask the electrician if the wiring and installation are inspected by a local electrical inspector. In some cases the insurance company will not insure a house unless it has been inspected and certified to meet the local codes. In some cases the National Electrical Code is specified as the bare minimum. The local code adds to this to meet special regional requirements.

It is very expensive to change plans once construction begins. The proper thing to do is go over the plans with a qualified electrician. Then tell the electrician your problems and what you want to do with each room before construction starts. This way the proper electrical system can be designed and installed.

To be realistic — no house is ever properly wired. The persons living in a house change in their requirements for electrical power. They buy new devices and they rearrange the furniture. In some instances, use of an extension cord can solve a problem. However, the use of extension cords should be limited. The proper wire size should always be used for a particular circuit's requirements.

As an electrician you should check the latest issue of the National Electrical Code for any new procedures or limitations. Keep current.

QUESTIONS

1. How do you figure circuit load?
2. Where are convenience outlets located?
3. Where are wall switches located?
4. How do you determine the number of lighting or convenience outlets needed in a room?
5. What is a dual-function outlet?
6. What is the symbol for a single-pole switch on a floor plan?
7. What do you have to consider when planning a house wiring plan?
8. What is a GFI?
9. Where are GFIs required?
10. What is the wall width needed to require a receptacle?
11. What type of special outlets should be installed in a bathroom?
12. How many wall switches must a bathroom have?
13. Where are receptacles required along a kitchen counter space?
14. What is an example of an appliance with a dedicated space?
15. How many outlets must you have for a laundry room?
16. How far away from a closet shelf must a light fixture be?
17. How many outlets must you have in a family room?
18. What are Code requirements for storage spaces?
19. How big must a porch be before it is required to have a lighting outlet?

CHAPTER 7

Equipment and Tools for Wiring

For an electrician to do the job properly, it is necessary to have the right tools. Tool requirements vary with the job. Some need only a minimum of tools. Others require some rather complicated devices.

The more complicated tools are used by the electrician who works on commercial and industrial installations. Some of these tools, which are very expensive, are bought by the electrical contractor. It is necessary, however, for an electrician to have used such tools, or at least to have knowledge of their existence.

This chapter will endeavor to show both the simple and the complicated tools that are used every day by an electrician. Some electricians have had to work with devices they improvised themselves. These have later become standard equipment for certain jobs. The ingenuity of electricians is well known to those people associated with them. Their ability to get a job done efficiently is also well known. Yet, most of this efficiency is due to the particular individual's skill in utilizing the equipment at hand.

BASIC TOOLS

Figure 7-1 illustrates the minimum tools necessary for house wiring or for adding an outlet or switch to existing wiring.

- *Hammer.* Although primarily for driving nails, staples, or for fastening hangers, a hammer may be used to anchor an object to concrete. Fig. 7-2. In Fig. 7-3 a hammer is being used to attach a box to a 2-inch × 4-inch stud. The hammer has many uses to an electrician.

- *Bit.* A drill bit may be used to drill holes in wood or soft metal. Some more extensive uses will be shown later in this chapter. The bit in Fig. 7-1 has been designed for use in a brace, for operation by hand.

- *Bit brace.* This tool is very useful in drilling holes for the location of boxes and wires, and for fishing wiring out of hard-to-reach locations. Fig. 7-1.

- *Keyhole saw.* The keyhole saw cuts circles and irregular shapes. Fig. 7-1.

- *Hacksaw.* Used for cutting cable (BX or Romex), the hacksaw may be used to cut plaster or laths, or any metallic conduit, thinwall or rigid. Fig. 7-1.

- *Test light.* A test light is used to trace circuits, test fuses, and test circuits to determine whether power has been applied. Fig. 7-1 and 7-4. It can also indicate the voltage and whether it is AC or DC.

- *Multi-purpose tool.* Used to strip wires and crimp connectors, this tool will also cut screws without ruining the threads. Fig. 7-1.

- *Folding rule.* This standard, 6-ft folding rule is in every electrician's tool kit. It is used to check the proper location of boxes, wiring, and fixtures. Fig. 7-1.

7-1. A typical tool kit for an electrician. (Sears)

A **Hammer.** For driving staples, nails, fastening hangers.

B **5/8 inch Bit or Drill.** For boring wood or soft metal.

C **Bit Brace.** For use with drill, screwdriver, reamer.

D **Keyhole Saw.** For cutting circles and irregular shapes.

E **Hacksaw.** For cutting cable, plaster or laths.

F **Test Light** to trace circuits, test fuses and lines.

G **Multi-purpose tool.** Cuts and skins wire, attaches terminals, etc.

H **6-foot Folding Rule.** For measuring wire, openings.

J **Chisel.** For notching studs, joists, plaster, flooring and lath.

K **Lever-jaw Wrench.** Use as plier, lock wrench, pipe wrench.

L **Linesmen's Pliers.** For gripping locknuts, cutting wires.

M **Wire Cutter, Stripper.**

N **Jackknife** for cutting insulation.

P **Fish Tape and Reel**—fish wire thru wall and conduit.

R **Conduit Bender.** A handy time-saver.

S **Screwdriver** to tighten screws, locknuts.

7-2. *Using a hammer to set an anchor.*

1.

2.

3.

7-3. *Nailing a box to a stud.*

7-4. *A circuit tester. (Ideal)*

7-5. *Pliers used to break off the fins on a light dimmer.*

● *Chisel.* A necessary tool for making the required small changes in the woodwork when mounting a fixture or box, a chisel may also be used to notch studs, joists, plaster, flooring, and lath. Fig. 7-1.

● *Lever-jaw wrench.* This vise grip can be used as a pliers, lock wrench, or pipe wrench. Fig. 7-1.

● *Side cutters (sometimes called linesmen's pliers).* This tool probably gets as much use as any other in the electrician's tool kit. It can be used for gripping locknuts, cutting wires, and breaking fins off dimmers. Fig. 7-5. It can be used to work on floor

7-6. *Pliers being used to twist new Romex connector.*

7-7. *A fish tape and reel.*

7-8. *Using fish wire. (Sears)*

Push 12-foot fish wire, hooked at two ends, through hole on 2nd floor. Pull one end out at switch outlet on 1st floor.

Next, push 20-25-foot fish wire, hooked at both ends thru ceiling outlet (arrows). Now fish until you touch the first wire.

Then withdraw either wire (arrows) until it hooks the other wire; then withdraw second wire until both hooks hook together.

Lastly, pull shorter wire thru switch outlet. When hook from long wire appears, attach cable and pull thru wall and ceiling.

7-9. *A Romex stripper, or ripper.*

7-10. *A wire cutter. (Ideal)*

boxes. This tool also serves as a crimper of new-type connectors for Romex, Fig. 7-6, and for removing twist-outs and knockouts.

• *Wire cutter, stripper.* A useful device for stripping wire, it is adjustable for the wire size being used. This tool makes it possible to strip insulation without cutting or marring the copper or aluminum conductor. It may be used to cut wires. Fig. 7-1

• *Jackknife.* A jackknife has many uses. The electrician cuts insulation with it without scoring the conductor. In addition, it might be used to cut tape. Fig. 7-1

• *Fish tape and reel.* This device, as the name suggests, is used to fish wire through walls and conduit. Figs. 7-7 and 7-8.

• *Conduit bender.* The conduit bender is handy for making bends needed for turns in the EMT (thinwall conduit) or rigid conduit. Some special tools have been designed and manufactured for the electrician working with both rigid and with the new plastic conduit. Fig. 7-1

• *Romex stripper, or ripper.* The Romex stripper, or ripper, (Fig. 7-9) is used to remove the thick coating of insulation from the two or three conductors in the cable. The ripper has a tip that sticks up inside the unit and will fit into the Romex insulation when the tool is squeezed. The tool is pulled to rip the outside coating of insulation. Then the insulation is cut with a pair of wire cutters, or with a diagonal cutter such as the one in Fig. 7-10.

• *Standard wire gage.* This gage checks the size of wire. Fig. 7-11. By placing a wire

7-11. *A wire gage. (Brown & Sharpe)*

7-13. *A handsaw.*

7-12. *Using a wire gage. Note: Wire thickness is measured by the slot, not the hole. (Brown & Sharpe)*

through the slot (not the hole) in the gage, it is possible to determine the size of the wire. Fig. 7-12.

• *Handsaw.* A handsaw is handy for cutting studs, or for making a box fit a bracket or fixture. Fig. 7-13.

• *Electric drill.* Electric drills have, in many cases, replaced the hand brace and bit. The drill shown in Fig. 7-14 is battery operated. It is very handy on new construction sites where power is not yet available.

• *Hole saw.* A number of hole saws are available. The saw in Fig. 7-15 first drills a pilot hole, then cuts a hole the diameter of the curved saw blade. In Fig. 7-16, you can see the different diameters available (only one saw blade is left in the tool when used).

• *Spin-tite wrenches and screwdrivers.* Spin-tite wrenches are screwdrivers with sockets on the end. Figures 7-17 and 7-18 show the variety of screwdrivers usually in-

7-14. *An electrician drilling a hole in a stud with a battery-operated electric drill. (Black & Decker)*

REPLACEABLE BLADE

TOOLHEAD

SHANK

PILOT DRILL

7-15. The parts of a hole saw. (Misener)

7-16. Hole saw blades. You take out the ones not being used.

cluded in an electrician's tool kit. Both the standard-type head and the Phillips head are included.

• *Chalk line.* This gadget comes in handy when a straight line is needed for a distance of more than a few feet. The string is covered with chalk as it is pulled from the case. When pulled tight and snapped, it produces a white line. Fig. 7-19.

• *Fuse puller.* A plastic-handled fuse puller is a tool used to safely extract fuses from industrial or commercial distribution boxes. Fig. 7-20.

7-17. Spin-tite socket wrenches.

7-18. Screwdrivers.

7-19. A chalk line.

7-20. A fuse puller. This is used for cartridge-type fuses.

7-21. An awl. It is used to start screws by making a pilot hole or for any number of other applications.

- *Awl.* The awl has a variety of uses. It can be used to mark lines or make holes, and as a tool for prying staples loose. Fig. 7-21.
- *Soldering irons and torches.* Soldering irons and guns are common tools for electronics technicians, and electricians also use them on occasion. However, if there is no electrical power available at a construction site, it becomes necessary to find tools that operate on other power. A propane torch is efficient here. It may be used for soldering either small or large jobs. Fig. 7-22.

There are other tools that the electrician can obtain for use on special jobs. Those listed here represent the necessities. A toolholder, shown in Fig. 7-23, comes in handy.

The offset needed to match up knockout holes in boxes to a piece of flat-mounted conduit can be accomplished with a special bender for $\frac{1}{2}$-inch and $\frac{3}{4}$-inch thinwall EMT.

7-22. A propane torch. (Bernzomatic)

7-23. An electrician's toolholder. (Eller)

7-24. Using a conduit bender to make a fit. (Greenlee)

7-25. A hand bender for EMT and rigid conduit. Note the guide marks on the head of the bender. (Greenlee)

Figs. 7-24 and 7-25. This bend in the conduit eliminates the need for connectors whenever the EMT is used in exposed locations, thus preventing a lot of trial-and-error bending and cutting. Fig. 7-26. Tools are available for all special jobs.

TOOLS FOR GENERAL USE

Saws and Chisels

• *Saws.* One of the most-used saws on the job is the *hacksaw.* Fig. 7-27. It comes in handy for cutting pipe, conduit, or any metal. A *compass saw* is used for cutting wood

7-26. A utility box surface-mounted with an offset bend. (Greenlee)

7-27. A hacksaw. (Greenlee)

7-28. A compass saw blade for woodwork. (Greenlee)

7-31. Firmer chisels. (Greenlee)

7-29. A flexsaw. (Greenlee)

7-32. A framing chisel. (Greenlee)

7-30. A utility chisel. (Greenlee)

to make way for conduit, or a box, or cabinet. Figure 7-28 shows a blade for a compass saw. The saw handle is similar to that of a hacksaw, and shaped to fit the hand.

The *flexsaw* is a compact saw for close-quarter jobs. Fig. 7-29. It uses a hacksaw blade, which flexes for a cut flush to the surface. It is used to cut metal.

Hacksaw blades suitable for the type of material to be cut should be chosen. They are available in 10-inch and 12-inch lengths. But the number of teeth per inch should depend on the shape and size of material being cut. The three teeth designations available are 18, 24, and 32 teeth per inch. The more teeth the better for thin material, or for small-diameter pipe.

• *Chisels.* Chisels are used to remove wood from studs or framing to allow clearance for mounting conduit or metal boxes. There are a number of chisels available for electricians or carpenters. Here are some usually chosen by electricians or line-workers.

The *utility chisel* in Fig. 7-30 is a special-purpose type used by utility and telephone line workers to cut daps in poles for crossarms. The blade is 2 inches wide and heat treated for hard wear and use.

The *firmer chisel,* with beveled edges, is made of high-carbon tool steel. Fig. 7-31.

The *framing chisel* can be used for heavy framing, or as a wrecking tool, or like the firmer chisel, to chip wood. Notice that it does not have beveled edges. This is one way to distinguish it from the utility chisel. Fig. 7-32.

The *gouge* is a woodworker's tool. The electrician uses it to remove excess wood for the installation of conduit. Fig. 7-33.

Drill Bits

Woodworking tools are valuable to an electrician who must make holes in wood to pull wire or place conduit. A number of drills are available for boring holes in wood. A hand-operated drill, called a *brace,* takes a number of drill bits suitable for each job. Fig. 7-34.

The bit brace uses the principle of the wheel and axle in creating a driving force. The jaws of the chuck hold the bit in the brace during operation. Pressure is applied to the *head* of the brace (the rounded, wooden knob on the end). A turning motion is applied by grasping the *handle* in the middle of the brace, and rotating it. Drill bits have a screw tip to catch the wood and pull the cutting edge forward. The *twist* of the bit causes smooth flow of wood chips to the outside of the hole.

Some braces have a ratchet wheel that allows for turning the bit in a clockwise direction only, but allows the *bow* (bent part of the brace) to be turned in the opposite direction while the bit is stationary in the wood. It is then possible to bore a hole in a piece of wood where the full swing of the bow is not possible. A *cam ring* can be turned to allow for ratchet action or for full operation.

An electrician should be able to identify the proper drill bit for a job. Some bits are more efficient than others due to their design and availability in a number of sizes. The sizes of the following bits are given in customary measurement. Drill bits are available in metric sizes. Eventually, such metric drill bits will be more commonly used.

- *Machine spur bit.* The bit in Fig. 7-35 is properly referred to as a *machine spur bit.* It is a flat, wood-boring power bit designed for use in electric drills, to bore holes in any wood, at any angle, and at any speed. It is usually a little over 6 inches in length and comes in cutter sizes ranging from $\frac{1}{4}$ inch to $1\frac{1}{2}$ inch, in increments of $\frac{1}{16}$ inch.

- The *multi-spur pipe bit* has three milled flats for use in portable electric or pneumatic drills, and in stationary boring machines. Fig. 7-36.

7-33. A gouge. (Greenlee)

7-35. A machine spur bit. (Greenlee)

7-34. A brace.

7-36. A multi-spur pipe bit. (Greenlee)

- *Plain-type expansive bits* for use in a brace have a clamp and screw for locking the cutter in place. Fig. 7-37. This bit may be obtained in $\frac{1}{2}$- to $1\frac{1}{2}$-inch, $\frac{7}{8}$- to $1\frac{3}{4}$-inch, and $1\frac{3}{4}$- to 3-inch cutters.

- *Auger bits.* A *bellhanger's drill* is used for telephone and other wiring installations. Sizes are $\frac{3}{16}$ to $\frac{3}{4}$ inch in $\frac{1}{16}$-inch increments. It comes in 12-, 18-, and 24-inch lengths. Fig. 7-38.

- The *uni-spur electrician's power bit* is a double-twist, smooth-boring bit used by electricians, plumbers, and pipe fitters. Fig. 7-39. It is especially designed for use in electric drills. The screw point feeds at sixteen turns to the inch. Note the two set-screw holes for use in a power bit extension or brace adapter. The overall length is $8\frac{1}{2}$ inches. It comes in $\frac{5}{8}$- to $1\frac{1}{4}$-inch size with $\frac{1}{16}$- and $\frac{1}{8}$-inch intervals, depending upon the diameter needed.

- The *single-twist power bit* is a smooth-boring bit designed for electric drills. Fig. 7-40. It has excellent chip clearance due to its single-twist design. It may be used to bore holes from $\frac{5}{8}$ to 1 inch with $\frac{1}{16}$-inch difference between various drill sizes. Overall length is from 5 to 7 inches, depending upon the drill size. With extensions, it will drill through floor joists, studs, or through existing construction for installation of Romex.

- A *solid-center power-car bit* has a solid center to add stiffness and strength. Fig. 7-41. It has a single twist for excellent chip removal from the cutting area. The screw point is of a medium pitch. It is designed for use in three-jaw chucks of $\frac{3}{8}$ inch and larger. It can be used with the bit extension. Overall length is about 16 inches, available in sizes from $\frac{3}{8}$ to 1 inch in sixteenths of an inch.

7-37. *An expansive bit. (Greenlee)*

7-41. *A solid center power-car bit. (Greenlee)*

7-38. *A bellhanger's drill bit. (Greenlee)*

7-42. *A power-pipe bit. (Greenlee)*

7-39. *A uni-spur power bit. (Greenlee)*

7-43. A single-spur impact-wrench bit. *(Greenlee)*

7-40. *A single-twist power bit. (Greenlee)*

7-44. *A rafting auger. (Greenlee)*

7-45. *A ship-auger power bit. (Greenlee)*

7-46. *A brace-bit extension. (Greenlee)*

7-47. *A power-bit extension. (Greenlee)*

- The *power-pipe bit* is ideal for boring holes in wood to allow passage of pipe, conduit, and tubing, and for other uses requiring large holes. Fig. 7-42. It is designed for rough, rugged boring. Lengths run from 14 to 17 inches overall, with cutter sizes available from $1\frac{1}{4}$-inch diameter for $\frac{3}{4}$-inch pipe, $1\frac{1}{2}$-inch diameter for 1-inch pipe, $1\frac{3}{4}$-inch diameter for $1\frac{1}{4}$-inch pipe, 2-inch diameter for $1\frac{1}{2}$-inch pipe, and $2\frac{1}{2}$-inch diameter for 2-inch pipe.

- An *impact-wrench bit* can be identified by the grooved portion near the end of the bit. Fig. 7-43. This single-spur (spur refers to the cutting edge near the screw tip) impact bit bores holes from $\frac{3}{8}$ to $1\frac{1}{16}$ inch in telephone or power line poles quickly, easily, and safely, using an impact wrench. Two sizes fit all impact wrenches with the use of $\frac{7}{16}$- or $\frac{5}{8}$-inch hexagonal adapters. Overall length is 12, 16, or 24 inches, depending upon the twist length (8, 12, or 18 inches) needed. The diameter of the drill can be $\frac{3}{8}$ to $1\frac{1}{16}$ inch, depending upon the bit chosen.

- The *rafting auger* is sometimes referred to as a *boom auger.* This tool has a double twist with a flat-cut pattern head. Fig. 7-44. It is available in overall lengths of 14 to 20 inch, with hole diameters ranging from $1\frac{1}{4}$ to 4 inches.

- A *ship-auger bit* has a shank for use with $\frac{3}{8}$ inch and larger electric drills. Electricians, plumbers, and other skilled workers employ it to drill deep holes or where extra reach is needed. Fig. 7-45. The ship auger has a medium-feed screw point and cutting head, with side lip to withstand rough usage. It can be used with a power bit extension to increase its length. Overall length is about 17 inches normally, and it can be had in sizes from $\frac{3}{8}$ to 1 inch.

- *Extensions:*

A *brace-bit extension,* Fig. 7-46, is used to extend the distance from the drill bit to the power source. It allows an electrician to get at places otherwise inaccessible because it allows for operation of the brace in a restricted area.

Power-bit extensions are needed to get at some locations where the power drill and normal drill bit could not. The bit extension in Fig. 7-47 has one flat side with two set-screw holes permitting two or more extensions to be locked together for longer reach. They come in either 18- or 24-inch lengths.

The *power-bit expansion* is an extension with a $\frac{1}{4}$-inch hex end to take $\frac{1}{4}$-inch flat *zip bits*. Fig. 7-48. The hex configuration makes for a secure bit fit in the extension. This one is available with an overall length of 12 inches.

The previous list is not by any means the extent of drill bits available. This has been an effort to illustrate some of the available types, and to show the difference between single-twist, double-twist, and solid-center bits. The spur side, or cutting edge, may make a difference in the job or the finish of the hole. The rate of feed and the ease with which a hole can be bored are also important with a hand-operated brace.

Extensions are basically of three types. They may vary by having either hand- or power-driven, or round- or hexagonal-shaped shanks.

Special Equipment

● *Nail puller.* The quick, easy removal of nails is important to an electrician. Wooden boxes, crates, or old construction frequently require nails removed so that the electrician can get the job done. Fig. 7-49.

The impact handle slides on an upright claw bar for quick hammer action, driving the claw under the nailhead. The claw hook is extended to form a leverage foot and to guide the handle for fast, safe positioning of claws, and to provide maximum pull on the nail.

Note: Table 7-A lists a few conversions that may prove helpful in converting some of the customary measurements in this chapter.

7-48. *A power-bit expansion for use with hex-shank bits. (Greenlee)*

7-49. *A nail puller. (Greenlee)*

Table 7-A. Metric Measurements*

1 inch	= 25.4 mm
1 foot	= 0.304 8 m
1 sq in	= 6.451 6 cm^2
1 sq ft	= 0.092 903 04 m^2
1 cu in	= 16.387 064 cm^3
1 cu ft	= 0.028 316 846 592 m^3

*These units are given here so that the measurements in this chapter can be converted to metrics if the occasion arises.

It may be that in the future all tools will be metric. This will mean that a number of things will have to change first. The fasteners and the materials worked with by tools will have to be converted to metric standards.

No standards exist at the moment, but it may not be too long. Until metric standards are established, the boxes and other equipment will continue to contain such requirements as "only 6 #14 conductors can be placed in a box with 12.5 cubic inches." The entire National Electrical Code Handbook will have to be converted to metrics before the standards are found to be applicable to those working in the field.

QUESTIONS

1. List the minimum tools needed for doing electrical work.
2. For what purposes is a jackknife used in the electrical trade?
3. What is a flexsaw? Where is it used?
4. What is the difference between a chisel and a gouge?
5. What is a brace?
6. What is an auger bit? Where would an electrician use one?
7. What use would an electrician have for a nail puller?
8. Electrical boxes are measured in cubic inches. Suppose you have one that measures 12.5 cubic inches, what would be the metric equivalent?

House Service and Circuits

For an electrical system to operate properly, it is necessary to design the system so that the proper amount of power is available. This requires the house to be wired so that wire of the proper size serves each plug in the house.

Most power is distributed locally within a neighborhood by overhead wires. Fig. 8-1. The wires are usually located in the rear of the house. In some communities the wires are buried and come into the house near the basement or foundation wall. Power is

8-1. *Stringing overhead lines for local service in a residential neighborhood. (CILCO)*

8-2. *A transformer on a pole located at rear of the house. Note the high-voltage line on top and the low-voltage (240-volt) line on the bottom, with a take-off for the house. The bottom line is the cable television coaxial cable.*

Twisted bare conductor

Insulated conductors

Bare conductor helically

8-3. *Note how the twisted bare conductor is made from the outside coating of the other two wires.*

TYPE SE –

8-4. *Type SE cable. This cable has three wires: one red, one black, and an uninsulated wire that forms a protection for the other two. The stranded, uninsulated wire is twisted at the end to make a connection. See Fig. 8-3.*

brought from the pole or transformer into the rear of the house by one black, one red, and one white (uninsulated) wire. Fig. 8-2. These wires may be three separate ones, or they may be twisted together to look like one cable. Fig. 8-3.

Once the cable is connected to the house, it is brought down to the meter by way of a sheathed cable with three wires. These are red, white, and black. The white wire is usually uninsulated and twisted as shown in Fig. 8-4. The uninsulated wire is the ground or neutral.

Figure 8-5 shows how the power is fed from the transformer to the distribution box.

SERVICE ENTRANCE

All services should be three-wire. The capacity of service entrance conductors, and the rating of service equipment, should not be less than that shown in Table 8-A. This is a quick way to determine the type of service required. There are more accurate ways. These will be explained later in this chapter.

These capacities are sufficient to provide power for the following:

8-5. Power from the transformer is fed through the wires and down the side of the house to the meter. Then it goes through the wall at a lower level to a distribution box in the basement.

Table 8-A. Minimum Service Capacities for Various House Sizes

Floor area		Minimum service capacity
m²	sq ft	
Up to 93	Up to 1000	125 amperes
93-186	1001-2000	150 amperes
186-279	2001-3000	200 amperes

- Lighting.
- Portable appliances.
- Equipment for which individual appliance circuits are required.
- Electric space heating of individual rooms.
- Air conditioning.

A larger service may be required for larger houses. It may also be required if a central furnace or central hydronic-boiler is used for electric space heating.

Many major appliances in the kitchen require individual equipment circuits. Thus, where practicable, electric service equipment should be located near or on a kitchen wall. This will minimize installation and wiring costs. In addition, such a location will often be convenient to the laundry. This will minimize circuit runs to laundry appliances.

Prepare Your Plans

The first step in wiring a house is planning. This means you should make accurate plans and know exactly how much material you will need. Once you have learned about the boxes, wire, and tools you can then take a close look at the Codes.

The local power company can advise you regarding the kind of service to use for your house. In most instances, the local power company will take the power right up to the house. The rest of the wiring is your responsibility.

The electrician works on a house at two of the construction phases. The electrician is needed to put in the wiring while the house is still in the roughed-in stage. This means a new house will have the wiring installed before the drywall goes onto the studs. Once the drywall workers and carpenters have finished their work, the electrician needs to return. At this time the electrician will install the plugs, receptacle faceplates, circuit breakers, chandeliers, and lights. Switches will be connected and circuits tested for proper operation. In some cases the electrician will have to install the furnace, the electric heating, or the electric range. Various stages of construction will make it obvious that certain wiring should be done in an area before it is closed up. This is where planning is very important. It is much easier to wire a house while it is still unplastered than when it is walled in and ready for occupancy.

Permits

Before starting the installation, check with your local power company to make sure what permits are necessary. Most permits specify that you will have someone inspect the installation after you have finished. This inspection will determine if the installation is safe and free from possible fire hazards.

Local Regulations

In some communities, local regulations supersede the National Electrical Code. Make sure you know what these local regulations are before starting the installation. Make sure the materials you use are approved by the local power company. If not, you may have trouble getting service once the job is done.

Some regulations apply as a minimum. Those provided as guides by the National Electrical Code relative to the stringing of feeder wires are similar to those shown in Fig. 8-6. Here, a roof with a rise greater than 4 inches in 1 foot would be difficult to walk on easily. Therefore, one of the exceptions to the Code will allow a clearance of 3

Roof slope not less than a 4-12 pitch (4 in. rise in 1 ft)

Required clearance 3 ft min.

Service drop not over 300 V

8-6. *Clearance required for a 4-12 pitch roof in the path of a power line.*

4 in.

12 in.

NOTE: Clearance is specified for drop conductors—not drip loop service entrance conductors

Required clearance from window or door on elevated porch

3 ft min.

Service drop

Service entrance conductors in cable or raceway

Meter

8-7. *Minimum clearances for windows and doors for the installation of a service drop.*

Service drop

Above window

Meter

feet between the roof and an overhead power line if the line does not have over 300 volts. Figure 8-6 illustrates such a situation.

The electrician's main concern is the actual service drop and its entry into the house. The location of the entrance is important. The Code does have something to say about where the service can be located. Look at Fig. 8-7. Note that the service is required to be at least 3 feet from a window.

8-8. *Proper and improper ways of installing a service head and service drop.*

The possibility of rot caused by water can be minimized by placing the service head as shown in Fig. 8-8. The installation of this service head is usually the responsibility of the residential electrician.

Service from Head to Box

The service wires are not connected to the power company's lines until the inside of the house is properly wired and inspected. Figure 8-9 shows how the service is connected and the meter socket is inserted in the line. The overhead connections can be made if the meter is left out of the socket. This causes the rest of the service to be inoperative until the meter is properly inserted in the socket by the power company.

3 lead wires
must extend out
of entrance
head at least
36 in. Middle wire
is the neutral bare
wire and usually
connects to center
wire in view above.

Metal strap

Watertight
connector

Meter
socket

Clamped to a cold
waterpipe. However,
in some areas the ground
wire should be connected
to a rod driven into the
ground at least 8 ft.

Non-watertight
connector

Neutral wire
Red wire
Black wire
(240-volt circuit)

MAIN

RANGE

8-9. *Connected service using a fuse box
and external ground. (Sears)*

8-10. *The usual method of grounding city and town systems. (Sears)*

8-12. *The approved REA method of grounding wire system with ground rod. (Sears)*

8-11. *The usual method of grounding city and town systems using conduit. (Sears).*

Note how the neutral wire is also connected to the *cold* water pipe inside the house. Figs. 8-10 and 8-11. In some cases the house does not have a cold water system furnished from a local community water source. If a well is used for water, the neutral wire must be grounded as shown in Fig. 8-12.

A conduit might be used for the installation of the service into the house. Here, three wires are used. One is black, one is white, and the other is red. The size of the

SERVICE ENTRANCE FITTINGS
FOR HEAVYWALL OR THINWALL

Threaded caps

Clamp type
cap

Service entrance
cable connector

Service entrance
cable fittings

Cable
straps

THREADED CAPS FOR HEAVYWALL (RIGID) INSTALLATION

Service
mast

Entrance
ell for
conduit

-or-

Sill
plate for
non-metallic
cable

1
2
3
4
5

15- or 20-
ampere
general-
purpose
circuits

8-13. *Service entrance.*
(Sears)

6
7

Two
20-ampere
kitchen-
appliance
circuits

8   20-ampere laundry appliance circuit
9   15-ampere circuit for fuel-fired furnace
10   240-volt central air conditioning circuit
12

11   20-ampere workshop circuit
13 –15 120-240-volt range circuit
14 –16 240-volt hot water heater circuit
17 –19 120-240-volt washer-dryer circuit
21   20-ampere circuit for dishwasher
22   20-ampere circuit for garbage dispenser

Service to barn or outbuilding should be
through a separate entrance panel
connected to power take-off lugs in
main entrance panel.

Outdoor light
150 watts

18 20

60-ampere
feed to
sub-panel

1-2
4-5
3
6

168

wires will vary according to the amount of current you will be needing. This will be covered later in this chapter.

Conduit is limited in the size of wire it can handle. For instance, a $\frac{3}{4}$-inch conduit can take three No. 8 wires. The $1\frac{1}{4}$-inch conduit can take three No. 2 wires, three No. 3 wires, three No. 4 wires, or three No. 6 wires. A $1\frac{1}{2}$-inch conduit can take three No. 1 wires. The 2-inch conduit will handle three No. 1/0 wires, three No. 2/0 wires, or three No. 3/0 wires. Therefore, you can see the importance of the requirements being known first. If you know the amount of current you will be needing, you can determine the size of the wire. Then you can figure out which size conduit to use to hold the wires safely.

Wires from the utility company's pole to your building are called the *service drop*. These are usually furnished by the company. These wires must be high enough to provide proper clearance above grade. They also must not come within 3 feet of doors, windows, fire escapes, or any opening. See Fig. 8-7. The structure to which the service drop wires are fastened must be sturdy to withstand the pull of ice and wind.

Installation of the Service

Figure 8-13 shows how the conduit for the service is connected to the head and to the meter socket. Note how it enters the house with a waterproof fitting and then goes unspliced into the distribution panel in the basement.

Circuits are run from the distribution panel to various parts of the house. Some examples of circuits are shown in Fig. 8-13. Note that here there are twenty-two circuits available. Figure 8-9 shows how the fuse-type box is served by the entrance wires.

Conduit is fastened to the wall of the building by clamps placed every 4 feet. Note that the conduit should use an entrance ell to turn the conduit into the house. Such an ell has two threaded openings corresponding to conduit size. Use an adapter to fasten the conduit into the threaded opening at the top of the ell. Into the lower opening, fasten a piece of conduit to run through the side of the house. Use a connector to attach the conduit to the electric service panel.

Inserting Wire into a Conduit

After the conduit is installed, you can push the wires through the top hub of the meter, through the conduit, and out of the service head. All three wires must extend at least 36 inches out of the service head. This will allow enough wire for connecting to the power lines. Then the wires are brought down from the meter to the entrance ell. Remove the ell cover and pull the wires through to the service panel inside the house. Use a white-covered insulated wire for the neutral in the conduit.

DISTRIBUTION PANELS

There are many types of distribution panels made for the home. One type has fuses that screw in. These fuses must be replaced when they are blown, or open. Fig. 8-14. Another type is the circuit breaker box. A circuit breaker can be reset if the device is tripped by an overload. This type is gaining in popularity for home use. Many people do not want to be bothered with looking for a new fuse every time one blows. The circuit breaker may be reset by pushing it to the "off" position and then to the "on" setting. If

8-14. *A distribution box. This is a four-circuit box with screw-in fuses. It is used for an add-on in some older systems.*

8-15. *A distribution box with circuit breakers. (Square D)*

EXAMPLE OF 100-AMPERE FUSELESS SERVICE ENTRANCE PANEL

100-ampere main breaker (shuts off all power)

40-ampere circuit (120- to 240-volt) for electric range.

30-ampere circuit (240-volt) for dryer, hot-water heater, central air conditioning, etc.

Four 15-ampere circuits for general-purpose lighting, television, vacuum cleaner.

Four 20-ampere circuits for kitchen and small appliances and power tools.

Space for four 120-volt circuits to be added for future loads as needed.

8-16. *An example of a fuseless service entrance panel capable of handling a 100-ampere service. (Sears)* it trips a second time, the circuit trouble is still present. It should be located and re-moved before resetting the breaker again. Fig. 8-15.

For a closer look at the fuse replacement or circuit breaker, take a look at Fig. 8-16. This is a 100-ampere fuseless service entrance panel. (Note: This size is almost obso-lete today.)

ROMEX CABLE

Romex cable is used to carry power from a distribution panel box to the individual outlets within the house. This nonmetallic sheathed cable has plastic insulation cover-ing the wires to insulate it from the environment. Some types of cable may be buried underground. Figures 8-17 and 8-18 show the types most commonly used in house wiring.

Most new homes are wired with 12/2, or No. 12 wire with two conductors. One of the wires is white and the other black. Inside the cable is an uninsulated ground wire of the same size, No. 12. In the past, the smallest wire size used in homes was No. 14, with 2 conductors (written on the cable as 14/2 with ground or 14/2WG). Single con-ductors may be used in conduits as stranded or solid conductors. Fig. 8-19. Figure 8-20 shows how the wire is cut to remove the insulation without scoring the metal part of the wire.

TYPE NM —12–2

8-17. *Romex cable, 12/2, for use in home cir-cuits.*

SUNLIGHT RESISTANT — TYPE UF UNDERGROUND

8-18. *Romex that can be used for underground wiring.*

8-19. *Single wire, stranded, with plastic coat-ing. This can be used in conduits. It also comes in a solid copper wire.*

Wrong way   Right way

8-20. *The right and wrong way to cut insula-tion from a piece of wire. (Sears)*

8-21. *Sizes of wires. Note the relationship between the size and number.*

WIRE SIZE

The larger the physical size of the wire, the smaller the number. For example, No. 14 is smaller than No. 12. Number 14 has been used for years to handle the 15-ampere home circuit. Today, the Code calls for No. 12 for a 15-ampere circuit with aluminum or copper-clad aluminum wire. This is a safety factor. Aluminum wire cannot handle as much current as copper.

Figure 8-21 shows how the size and numbers relate in determining wire diameter.

Table 8-B lists the capacities of wires needed for various household equipment. Table 8-C shows conductor capacities for other household equipment. You should consider this information whenever you are planning an installation.

PLANNING THE RIGHT SIZE SERVICE AND CIRCUITS

Most houses have either a 150-ampere or a 200-ampere service from the pole in the back of the house to the circuit breaker box in the basement. Number 1/0 or No. 3/0 (type RHW insulation) three-wire is usually used. This usually provides sufficient power for household needs. One way to find out how much power is needed is to list the devices that use power in a house and how much. Add up the requirements. Divide the wattage by 120 to find out how many amperes would be needed at any time.

In a house with an electric range, water heater, high-speed dryer or central air conditioning, together with lighting and the usual small appliances, there is at least a 150-ampere requirement. If the house also has electric heating, a 200-ampere line should be installed.

Table 8-B. Conductor Capacities for Household Equipment

Item	Conductor capacity
Range (Up to 21-kW rating) *or*	50 A-3 W-115/230 V
{ Built-In Oven	30 A-3 W-115/230 V
{ Built-In Surface Units	30 A-3 W-115/230 V
Combination Washer-Dryer *or*	40 A-3 W-115/230 V
Electric Clothes Dryer	30 A-3 W-115/230 V
Fossil-Fuel-Fired Heating Equipment (if installed)	15 A or 20 A-2 W-115 V
Dishwasher and Waste Disposer (if necessary plumbing is installed)	20 A-3 W-115/230 V
Water Heater (if installed)	Consult Local Utility

Table 8-C. Conductor Capacities for Other Household Equipment

Item	Conductor capacity
Room Air Conditioners *or*	20 A-2 W-230 V
Central Air-Conditioning Unit *or*	40 A or 50 A-2 W-230 V
Attic Fan	20 A-2 W-115 V (Switched)
Food Freezer	20 A-2 W-115 or 230 V
Water Pump (where used)	20 A-2 W-115 or 230 V
Bathroom Heater	20 A-2 W-115 or 230 V
Workshop or Bench	20 A-3 W-115/230 V

150-Ampere Service

A 150-ampere service will provide sufficient electrical power for lighting and portable appliances. This would include, for example, a roaster, rotisserie, refrigerator, and clothes iron. If the dryer does not draw more than 8700 watts, the range no more than 12 kilowatts, and the air conditioner no more than 5000 watts, then 150 amperes will be sufficient. Do not add any more than 5500 watts of appliances, however. Table 8D shows how much power is required to operate small devices.

Table 8-D. Appliances and Their Wattages

Device	Wattage	Voltage	Current amps
	Small Appliances		
Blender	950	120	7.92
Clothes iron	1100	120	9.17
Electric fryer (small)	1200	120	10.00
Food processor	400	120	3.34
French frier	900	120	7.50
Hair dryer	1200	120	10.00
Hand drill (1/3 HP)	324	120	2.70
Microwave oven	1680	120	14.00
Mixer	165	120	1.38
Toaster	1200	120	10.00
Vacuum cleaner (1 HP)	746	120	6.22
Waffle grill	1400	120	11.67
	Large Appliances		
Clothes dryer	5000	240	20.83
Dishwasher	1200	120	10.00
Laundry circuit	3000	240	12.50
Range	12 000	240	50.00
Space heating	9000	240	37.50
Water heater	2500	240	10.40

8-22. *To add a wall switch for control of a ceiling light at the end of the run. (Sears)*

8-24. *To add a wall switch to control a ceiling light in the middle of a run. (Sears)*

8-23. *To add a new convenience outlet from an existing junction box. (Sears)*

8-25. *To add new convenience outlets beyond old outlets. (Sears)*

Branch Circuits

General-purpose circuits should supply all lighting and all convenience outlets, except those in the kitchen, dining room (or dining areas of other rooms), and laundry rooms. General-purpose circuits should be provided on the basis of one 20-ampere circuit for not more than every 500 square feet or one 15-ampere circuit for not more than every 375 square feet of floor area. Outlets supplied by these circuits should be divided equally among the circuits. Figs. 8-22 through 8-30.

LOCATING THE COMMON TERMINAL

The common terminal is the point where the *hot* wire (to be switched) is connected. This allows the switching action to complete the circuit by contacting one of the "traveler" terminals.

Study

The location of the common terminal on a three-way switch may vary according to the manufacturer. If the common terminal is not marked, use an ohmmeter to locate it. Turn on the ohmmeter. Place one lead on what you think is the common terminal. Place the other lead on another terminal. If the ohmmeter registers continuity (zero) with this arrangement, flip the switch handle or toggle. Then, the reading should be infinity (open). Leave the first lead on what you think is the common terminal. Move the second lead to the other switch terminal. If you read zero now that you have moved the switch handle or toggle, you have located the common terminal. However, if you read infinite (open) on the terminal, you have *not* located the common terminal. Then, you must move the first ohmmeter lead to another assumed common terminal and repeat the operation until you locate the common terminal.

8-26. To add a switch and convenience outlet in one box, beyond existing ceiling light. (Sears)

8-27. To install two ceiling lights on same line; one controlled by switch. (Sears)

8-28. To add a switch and convenience outlet beyond existing ceiling light. (Sears)

8-29. To install one ceiling outlet and two new switch outlets from existing ceiling outlet. (Sears)

8-30. Two methods of attaching ground wire to metal box.

These requirements for general-purpose branch circuits take into consideration the provision in the current edition of the National Electrical Code. Floor area designations are in keeping with present-day usage of such circuits. Figs. 8-31 through 8-39 (Pages 177-180).

For a close look at the internal wiring of a house, before it is covered with drywall or plaster, see Figs. 8-40 through 8-45 (Pages 180-182). Wiring of switches and receptacles, and locations within the circuits, are shown in Figs. 8-46 through 8-56 (Pages 183-185).

8-31. Here, the three-way switches control a single lamp. Feed is through the center-lamp box. The terminals marked A and B are the light-colored points to which red and white wires must be connected. Terminal C is the dark-colored (brass screw) point to which the black wire must be connected. (Sears)

8-32. Two three-way switches are tied together with a three-wire Romex cable. The lamp and switch box are fed with a two-wire cable. The terminals marked A and B are the light-colored screws to which red and white wires must be connected. Terminal C is the dark-colored (brass-colored screw) point to which the black wire must be connected. (Sears)

8-33. *Switching arrangements for single-pole (B), three-way (C), and four-way switches (D). Note the switches and their terminal screws. Also, note the terminals on the receptacle (A).*

8-34. *The four-way switch is in the middle. There is a three-way switch on the left and a three-way switch on the right. Two three-way switches and one four-way switch make it possible to control a lamp from three locations. Terminals marked A and B are the light-colored screws to which red and white wires must be connected. Terminal C is the dark-colored screw to which the black wire must be connected. Terminals AA show where the two ends of the red wire are connected between the four-way switch and the three-way switch on the left. Terminals BB show where the two ends of the white wire are connected to the four-way switch and the three-way switch on the left. (Sears)*

8-35. *Three wires feeding a single receptacle. Switched lower outlet—top outlet hot all the time.*

8-36. *Three wires feeding a single receptacle.*

179

8-37. A breakaway connection between the two outlets allows the red and black hot wires to be connected to the same duplex outlet.

8-40. The cable in this switch box is fed from the top. It is then brought out from the top and stapled. Next, it leads down and is stapled again below the box. Since this is an uninsulated box, it has no clamp or connector. The staple above the box serves as a strain relief.

this is a completed circuit

8-38. Schematic drawing of two three-way switches controlling a single lamp.

this is a incomplete circuit

crossing wires not connected

8-39. Schematic diagram of five locations to control a single lamp. Note there are never more than two three-way switches. Each time another location for control is added, a four-way switch is used.

8-41. *Location of an outlet where there is no insulation. This is the basement entrance and the switch box is located on the top right of the photo. It will serve to turn the basement light on and off, for light on the steps. Note how the cables have been stapled about 6 to 8 inches above the box and about 36 inches above that. The cable is held approximately in the middle of the stud.*

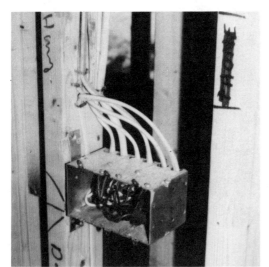

8-42. *Three-ganged switch boxes. Note how the wires are stapled to anchor them and how each box has a cable entering and leaving. This will be a three-switch control center.*

8-43. Note that the Romex feeds the box. BX (cable sheathed in metallic armor) is used to protect the asbestos-covered wires from the box to the lamp socket.

8-44. Note that the studs are drilled and the wires are fed to the switch boxes.

8-45. Wires left through the siding of a new house, to be attached to outlet boxes later. The boxes are covered at present. They will have the siding removed and connections made to the lamp fixtures and then covered and secured, making it a waterproof box.

QUICKWIRE HOLE
easy way to create a loop.

8-48. Note how the wire is released by pressing with a screwdriver blade into the slot. (Slater)

SPEEDWIRE
(Copper Wire Only)

8-46. Duplex receptacles being wired. A speedwire connection means that the wire is pushed into the hole to make contact. (Slater)

COMMON FEED ONLY

When plug is inserted into receptacle, pilot light automatically lights up.

8-49. Receptacle with pilot light. (Slater)

8-50. Switch and pilot light. (Slater)

WIRED AS PILOT LIGHT

When 3-Way switch is ON, pilot light glows. This circuit requires a ground wire for this operation.

8-47. Recessed outlet mounted in a 4-inch square box. (Slater)

WIRED AS NIGHT LIGHT (Breakoff Tab Removed)

When 3-Way switch is OFF, indicator light glows.

WIRED AS NIGHT LIGHT

When switch is OFF, indicator light glows.

Grounded power outlet controlled from 3-Way switch and 3-Way switch at another location.

WIRED AS PILOT LIGHT

When switch is ON, pilot light glows.

COMMON FEED

3-Way switch controls light only. Power outlet for grounded appliances.

8-51. *Switch and pilot light device wired with a three-way switch. (Slater)*

8-52. *Power outlet and switch can be used in a number of configurations for circuit control. (Slater)*

It is recommended that separate branch circuits be provided for both lighting and convenience outlets in living rooms and bedrooms. It is also recommended that the branch circuits servicing convenience outlets in these rooms be of the three-wire type, equipped with split-wired receptacles.

Space Heating and Air Conditioning Circuits

ELECTRIC SPACE HEATING. The capacity required for electric space heating should be determined from Table 8-E. The table shows maximum winter heat loss, based on the total square feet of living space in the home.

8-53. *Switch and outlet with a number of arrangements. (Slater)*

COMMON FEED

S. P. switch controls light only. Power outlet for grounded appliances.

Grounded power outlet controlled by S.P. switch.

SEPARATED FEED (Breakoff Tab Removed)

S. P. switch and grounded power outlet on separate circuits.

COMMON FEED

Two S.P. switches on same circuit. Each switch controls an independent light.

SEPARATED FEED
(Breakoff Tab Removed)

Two S.P. switches on separate circuit. Each switch controls an independent light.

8-54. *Two switches mounted in one unit with possibilities for controlling two loads in different ways. (Slater)*

WIRED AS NIGHT LIGHT

When switch is OFF, indicator light glows.

WIRED AS PILOT LIGHT

When switch is ON pilot light glows.

8-55. *Switch and night-light or pilot light with an addition of an external jumper. (Slater)*

8-56. *Baseboard electric heater. (Emerson)*

Table 8-E. Maximum Heat Loss Values*

(Based on Infiltration Rate of ¾ Air Change/Hr)

	Watts		Btuh	
Degree days	m²	sq ft	m²	sq ft
Over 8000	0.98	10.6	3.3	36
7001 to 8000	0.93	10.0	3.2	34
6001 to 7000	0.88	9.5	3.0	32
4501 to 6000	0.85	9.2	2.9	31
3001 to 4500	0.83	8.9	2.8	30
3000 and under	0.78	8.4	2.7	29

*For new homes. May be exceeded in converting existing homes.

8-57. *Electric heater showing thermostat. (Emerson)*

If electric space heating is installed initially, wiring should be as follows:

• For a central furnace, boiler, or heat pump: a three-wire, 115/230-volt feeder, sized to the installation.

• For individual room units: ceiling cable, or panels, sufficient 15-, 20-, or 30-ampere, two-wire, 230-volt circuits to supply the heating units in groups, or individually. Figs. 8-56, 8-57, and 8-58.

ELECTRIC AIR CONDITIONING. The capacity required for electric air conditioning is to be determined from the chart of maximum allowable summer heat gain, based on the total square feet of living space in the home. See Table 8-F.

If electric air conditioning is installed initially, wiring should be provided for the following:

• A heat pump, providing both winter heat and summer air conditioning; a three-wire, 115/230-volt feeder, sized to the installation.

• Individual room air-conditioning units, sufficient 15-, 20-, or 30-ampere, three-wire, 230-volt circuits to supply all units and a 20-ampere, 230-volt, three-wire outlet in each room, on an outside wall and convenient to a window.

8-58. *Smooth finish, flush installation of an electric heater. (Emerson)*

Table 8-F. Maximum Allowable Heat Gain (Btuh)*

Area to be conditioned		Design, dry-bulb temperature**			
m²	sq ft	32° C [90° F.]	35° C [95° F.]	38° C [100° F.]	40° C [105° F.]
70	750	15 750	18 000	19 500	21 000
93	1000	20 500	23 000	24 750	26 500
139	1500	27 000	30 000	31 500	33 000
186	2000	36 000	40 000	42 000	44 000
232	2500	45 000	50 000	52 500	55 000
279	3000	54 000	60 000	63 000	66 000

*Based on FHA minimum property standards.
**Based on dry-bulb temperature of less than 32° C [90° F.]; use 32° C [90° F.] values. For designed dry-bulb temperature exceeding 40° C [105° F.], use 40° C [105° F.] values. (To find °C, simply subtract 32 from °F and multiply by $5/9$.)

In some cases, neither electric space-heating nor electric air-conditioning equipment is installed initially. In such an instance, the service entrance conductors and service equipment should have the capacity required, in accordance with the appropriate chart. Spare feeder or circuit equipment should be provided, or provision for these should be made in panelboard bus space and capacity.

The plan should allow bus space and capacity for a feeder position that can serve a central electric heating or air-conditioning plant directly, or can supply a separate panelboard to be installed later. The panelboard would control circuits to individual room heating or air-conditioning units. Space heating and air conditioning are considered dissimilar and non-coincident loads. Service and feeder capacity need be provided only for the larger load, not for both.

SPACE-HEATING AND AIR-CONDITIONING OUTLETS. Since many different systems and types of equipment are available for both space heating and air conditioning, it is impractical to show outlet requirements for these uses in each room. Electrical plans and specifications for the house should indicate the type of system to be supplied and the location of each outlet.

Entrance Signals

Entrance push buttons should be installed at each commonly used entrance door and connected to the door chime. They should give a distinctive signal for both front and rear entrances. Electrical supply for entrance signals should be obtained from an adequate bell-ringing or chime transformer.

A voice intercommunications system permits the resident and a caller to converse without opening the door. This is convenient and is an added protection. It may be designed for this purpose alone, or it may be part of an overall intercommunications system for an entire house.

In a smaller home, the door chime is often installed in the kitchen, providing it will be heard throughout the home. If not, the chime should be installed at a more central location, usually the entrance hall. In a larger home, a second chime is often needed. This will ensure that it is heard throughout the living quarters.

Entrance-signal conductors should be no smaller than the equivalent of a No. 18 AWG copper wire. (AWG is the abbreviation for American Wire Gage.)

Communications

An intercommunications system, with a station in each room of the living quarters, is recommended. It may be arranged for voice communicaton to the entire house, or selectively to an individual room only. Music from AM or FM radio, stereo, or recordings may be fed into the system. It may also be used to communicate with a caller without opening the door.

For the larger home with accommodations for a resident servant, a dining room-to-kitchen signal is also recommended. This can be operated by a push button attached to the underside of the dining table or by a floor tread placed under the table. In some cases a built-in annuniciator in the kitchen, with pushbutton stations in various rooms of the living quarters, or a separate intercommunication telephone system is desirable.

Intercommunication systems and telephones should be operated from a power unit recommended by the manufacturer of the system. They should be supplied with a minimum of 115 volts.

For telephones, it is recommended that raceways and outlets be installed at the time of construction. Outlets, if desired, should be located in the kitchen, living room, bedrooms, and other rooms as requested. The telephone company installs concealed wiring in residential units during construction, on a selective basis.

It is suggested that the local telephone company be consulted for details of service prior to construction. They should be consulted particularly in regard to the installation of raceways and the location of protector equipment.

Television

An outdoor television antenna and lead-in connections require a nonmetallic outlet box at a convenient location. This is needed for connection of the master antenna system to the various other locations in the house. Allow for a 115-volt outlet near the antenna rotor control. It is more convenient if the antenna lead-in wire and the control circuits for the rotor are installed before the house has been closed in by drywall or by plaster. The appearance is also improved. Follow the rules and regulations for installing lead-in wire:

- Staples are to be used to hold only the insulated type of lead in place. Do not use staples on twin-lead.
- No right-angle turns.
- Do not run the antenna lead-in cable parallel to 60-hertz house current lines.
- Use an uninsulated box for making the termination connections to the plug-in receptacle. Fig. 8-59.

Radio

In their modern form, radio receivers used purely for the reception of broadcast communication seldom require antenna and ground connections. If FM reception is desired, it is recommended that provisions similar to those for television be included.

8-59. *TV outlets wired into boxes with receptacle for 120 volts to power the rotor. (Slater)*

Household Warning Systems

An automatic fire warning system is recommended, including installation of both smoke and heat detectors. A smoke detector should be located in the immediate vicinity of, but outside, the bedrooms. Additional smoke detectors placed in strategic locations around the home, and in each bedroom, are recommended. Heat detectors should be provided in each room and each hall of the living quarters. They should also be provided in the attic, furnace room, utility room, basement, attached garage, and each closet and partitioned-off storage area. Fig. 8-60.

The fire alarm bell should be clearly audible in all bedrooms, even with doors closed. An outside alarm bell is also recommended. A test button should be installed in the master bedroom or other desirable location.

The primary power supply should be 115 volts, feeding a separate and independent transformer recommended by the manufacturer of the system. A "power-on" visible indicator should be provided. A battery-operated standby source of power is also desirable.

Figure 8-61 pictures the newest type of fire-heat detector. It is mounted on the ceiling. It can detect heat and set off an alarm before a fire gets out of control. If the temperature rises above 57°C [135° F.], the fire-heat detector automatically activates an alarm. It should be used in living areas, bedrooms, and closets. It is only $1\frac{3}{4}$ inches in diameter and has a depth of $\frac{3}{4}$ inch. Another model, which looks exactly like the 57°C [135° F.] model, is available. It activates only at 93°C [200° F.]. This is designed for use in attics and furnace rooms, primarily.

A number of different ways are needed to trigger an alarm in case of an illegal entry. These require different switches. Some are used to detect the opening of a window. Others are connected to the door. And there is a mat with a switch to detect entry through a door. Figure 8-62 shows some of the different types of switches available to work in conjunction with a burglar alarm system.

8-60. *Wiring diagrams for home burglar alarm, smoke detector, and associated devices. (NuTone)*

8-61. *A fire-heat detector mounted on the ceiling tile. (NuTone)*

8-63. *An outside alarm bell. (NuTone)*

An outside bell is needed to alert neighbors. Fig. 8-63. They can call the police in case you are not at home. This bell can also be wired to the fire-smoke detector system and will awaken you in case of fire or smoke.

Sometimes a light and a siren are necessary. In this case, a model like the one in Fig. 8-64 will fill the order. It has a flashing red light as well as an audible warning.

A number of smoke detector systems are now on the market. Fig. 8-65. Most cities are requiring all buildings to have smoke detector systems. This will mean much work for electricians. When properly working, the systems can save many lives and millions of dollars each year. Many new systems will be coming on the market. It will be to your benefit to keep close check on the new methods used to detect smoke. The systems are basically simple. Try to obtain a wiring diagram of each system so that you can repair one if necessary. Most of the present-day smoke alarm detector units use a pilot

8-62. *Burglar alarm switches. (Nu-Tone)*

8-64. *A siren and flashing light warning device. (NuTone)*

8-65. *A smoke detector. (NuTone)*

light that must be on at all times. There is a manufacturer who uses *nuclear* materials in the smoke detector unit. Get the proper information before trying to check a unit. It may be dangerous to your health. Follow instructions properly and to the letter.

Installing Door Bells and Chimes

The installation of door bells and chimes is an easy job. The wire used is No. 18 wire. The connections are simple. Usually the wire comes in two or three-conductor cable. This is taken from the transformer to the switch and the chime as is shown in Fig. 8-66. The other side of the transformer goes to the 120-volt line. Usually the transformer

8-66. *Installation of door bells and chimes. (Sears)*

8-67. A door bell or chime transformer. Note the screw near the curled wires. This screw and the ears sticking up to hold the screw fit into a knockout in the entrance panel. They tighten to hold the transformer securely against the outside of the panel.

is mounted on the side of the circuit breaker box. The primary wires are fed through a hole in the box. The transformer has a screw that makes it easily secured to the box. See Fig. 8-67.

KILOWATT-HOUR METERS

Meters measure power used by a household, business, or industry. There are several types of kilowatt-hour meters. Figs. 8-68 and 8-69. Kilo means one thousand. Kilowatt-hour means 1000 watts of energy used for one hour. The unit of time for this type of meter is the hour. This means that whenever the meter reader records the meter

8-68. A kilowatt-hour meter. (Sangamo)

8-69. A three-phase kilowatt-hour meter. This type is designed for industrial use.

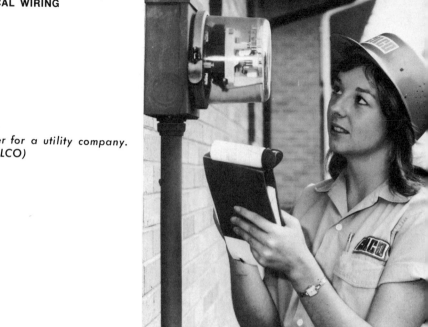

8-70. *A meter reader for a utility company.*
(CILCO)

reading—it may be a month from one reading to the other—the entire period in which power has been used is equated to one hour. The energy consumed is based on one hour. The number of watts would, in most cases, be thousands. Thus, the kilowatt-hour is used for measuring the power consumed, no matter what the length of time.

Sometimes the meter reader makes the rounds every twenty-eight days and sometimes every thirty-one days. Fig. 8-70. This means the reading is higher or lower in terms of usage. However, the time period for basing the consumption is still one hour. In a regular wattmeter, where there is no accumulative factor included, the reading and the watts consumed are in terms of the second, or what is being read at the moment the meter is used. This type of meter reading is usually done in the laboratory, or under test conditions where a definite consumption rate is necessary for a device.

8-71. *The parts of a kilowatt-hour meter and its hookup.*

8-72. *Read the dials. See if you obtain 7432 kilowatt-hours.*

In Fig. 8-71, the meter movement is composed of two coils. One coil is across the line to check the voltage. The other coil is in series with the load to check the current being drawn. This arrangement accounts for the voltage times the current, converting to watts, or power consumed. $P = E \times I$, or power is equal to the product of the voltage and current.

Reading a Kilowatt-Hour Meter

Take a look at Fig. 8-72 and read the amount of power consumed on the meter. If the dial pointer has just passed a number, read that number and not the next higher one. For example, in Fig. 8-72, the reading is 7432 kilowatts. Read from left to right and write down the numbers.

QUESTIONS

1. How is power usually distributed within a neighborhood?
2. What is a service entrance or service drop?
3. Why is an entrance service usually located near a kitchen?
4. Why are permits required for electrical work?
5. How often is conduit fastened to the wall of the building when it is used as part of the service for the house?
6. How is wire placed into a conduit?
7. What is a circuit breaker?
8. What is Romex cable?
9. What size wire is used for wiring houses with Romex?
10. The larger the physical size of the wire the larger the diameter of the wire. True or false?
11. What is the size of the wire usually strung from the pole in back of the house to the meter or circuit breaker panel?
12. What is the smallest size service, in amperes, that should be installed in a house today?
13. What is a branch circuit? Give an example.
14. Why would you want separate branch circuits for lighting and convenience outlets in a bedroom?
15. What is meant by the statement that central electric heating and air conditioning are non-coincident loads?
16. What size wire is used for signal communications such as a doorbell?
17. What is a kilowatt-hour?

CHAPTER 9

Installing Romex

ROMEX

Romex cable of the No. 14 or No. 12 size is usually employed in the wiring of homes, light industries, and businesses. To utilize safely the nonmetallic sheathed cable in a manner consistent with accepted wiring practices, it is necessary to use the switch and outlet box for terminations and connections. Splices are not allowed along the run of the cable unless the splice is housed in an appropriate box and totally enclosed with a cover plate.

When insulating material is used to make boxes and other equipment for wiring a house, it is not necessary to use a clamp or connector to hold the wire in place. Such a box is permitted when the wire or cable is supported within 8 inches of the box. Cable enters the box by way of a knockout. All knockouts should be closed if not used for cable.

Box Volume

The volume of the box is the determining element in the number of conductors that can be allowed in the box. For instance, a No. 14 wire needs 2 cubic inches for each conductor. Therefore, if a two-wire Romex cable of No. 14 conductors is specified, it means that a ground wire (uninsulated) is also included which will count as a conductor, and the three wires, or conductors, will require 6 cubic inches of space. A box used for this installation should have at least this amount. A 3- × 2- × 1½-inch device box has only 9.0 cubic inches and will be allowed to contain three of either No. 14, No. 12, or No. 10 conductors, but only two No. 8 conductors. Boxes manufactured recently will have the cubic-inch capacity stamped on them. You should check the National Electrical Code handbook for the number of wires allowed until you are familiar with the standards for boxes.

9-1. *An insulated metallic grounding bushing. (Union)*

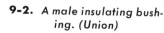

9-2. *A male insulating bushing. (Union)*

196

9-3. A nonmetallic switch box for a house. (Union)

9-4. A nonmetallic switch box. It will hold nine conductors of No. 14 wire, eight conductors of No. 12 wire, or seven conductors of No. 10 wire. (Union)

In Figs. 9-1 through 9-14 you will find a representative sampling of devices made of insulating material that are acceptable in house wiring. Read the captions under each figure and identify the characteristics for future use in wiring buildings.

BX

BX cable, flexible metal-covered cable, is covered in detail in Chapter 11. Here, it is discussed briefly as it compares to the use of Romex in circuits for homes.

9-5. A nonmetallic switch box, with nails used to mount it. It will hold seven conductors of No. 14; six of No. 12 or No. 10. (Union)

9-6. A nonmetallic switch box, designed with a bracket to mount to the wood or steel studs with SST tool. The box will hold fifteen No. 14 conductors, thirteen No. 12, or twelve No. 10 conductors. (Union)

9-7. A nonmetallic switch box. It will hold four devices. The bracket will fit between two studs located on 16-inch centers. The box will hold twenty No. 14 conductors, seventeen No. 12, or fifteen No. 10 conductors. Remember—the ground wire counts as one conductor. (Union)

9-8. *A cover for an outlet box or handy box for a duplex receptacle, nonmetallic. (Union)*

9-9. *A nonmetallic cover for a toggle switch. (Union)*

9-10. *A nonmetallic surface box. It will hold only three No. 14 conductors, and two No. 12 or No. 10 conductors. (Union)*

● *Connectors.* BX requires connectors that make a clamping action that holds the protective coating rigidly in place. They also make a good electrical connection for grounding purposes. The design of these connectors will vary with different manufacturers. However, the group of connectors shown in Figs. 9-15 to 9-20 will represent those made by a particular manufacturer. They will serve here as examples of connectors available for switch boxes, utility boxes, and both square and octagonal outlet boxes.

9-11. *A nonmetallic conduit box, round, with four $\frac{1}{2}''$ threaded knockouts. Suitable for fixture mounting, it has 12.5 cubic inches of space and a threaded knockout. It is also available with $\frac{3}{4}$-inch threaded knockouts. (Union)*

9-12. *A duplex receptable cover (REC), with gasket and stainless steel screws. It may also be obtained with a single receptacle hole. Note the REC on the outside to identify the outlet. (Union)*

9-13. *A ground continuity tester for use on energized circuits only. It is equipped with a ground and a neutral blade. If the light comes on when it is plugged into a grounding-type receptacle, the ground continuity is complete. If the light does not come on, there is a fault in the circuit. It uses a penlite cell for power. (Union)*

9-14. A box finder and cable tracer used to locate boxes or cables that are hidden from view. To use, plug it into one of the receptacles on the covered box circuit or attach it to one of the circuit wires. The transmitter will produce a signal. Tune in a small transistor radio to pick up the transmitted signal. Follow the cable by checking sound build-up or fade. As the box is approached, sound builds up due to more wire being folded back in the box. The box will be located at maximum signal strength. The unit is more efficient if the circuit ground wire is disconnected. It uses a 9-volt transistor battery. (Union)

9-15. A box connector for entrance service, for nonmetallic sheath cable or nonmetallic flexible tubing. The connector does not make a weatherproof connection to the box. (Appleton)

9-16. A Romex box connector designed for nonmetallic sheathed cable (Romex) or nonmetallic flexible tubing. It will accomodate 14/2, 14/3, 12/2, 12/3, or 10/2 cable. (Romex for home use is usually 14/2 or two wires of No. 14. Ground wire is assumed to be there, but designation will usually be 14/2 WG). (Appleton)

9-17. A BX box connector. Armored cable, or $\frac{3}{8}$-inch flexible steel conduit, may be accomodated with this connector. It can handle 14/3, 14/4, 12/2, 12/4, or $\frac{1}{2}$-inch cable. (Appleton)

9-18. A BX box connector. A duplex connector for armored cable, and for $\frac{3}{8}$-inch flexible tubing. (Appleton)

9-19. A BX box connector, with two screws for $2\frac{1}{2}$-inch flexible steel conduit. (Appleton)

9-20. A 90° BX box connector, used also for $\frac{1}{2}$-inch flexible steel conduit. It will handle 8/2, 8/3, 8/2 WG cable. (Appleton)

9-21. *(A) Bushing inserted into the BX to prevent damage to the insulation of the wire. (B) BX connector is slipped into place over the bushing and metal armor, and the screw tightened to hold the connector in place. (C) BX connector is held in place through a knockout in the box, by use of a locknut. (Sears)*

A *bushing* is needed to protect the wires inside the armored cable from abrasion and from cuts in the insulation. Note that the connector is mounted in the box for a positive holding action. Figure 9-21 shows a fiber bushing inserted in a BX.

Figure 9-22 shows how a *connector* is used to hold the Romex cable in place. *Note how the grounding clip is attached to the metal box.* Note also that part of the Romex insulation is extended through the connector into the box.

A number of sizes and configurations are available to the electrician for use as receptacles. A sampling of the different configurations available for installation in switch boxes and outlet boxes, both metallic and nonmetallic, is shown in Fig. 9-23.

Figures 9-24 through 9-35 (Fig. 9-35 on Page 204) show the different types of boxes available for use as switch boxes. Some installation procedures are also depicted.

9-22. *A Romex connector attached to a box. Note the ground clip for the uninsulated ground wire. (Sears)*

20 A, 125 V, Three wire, grounded, two pole

15 A, 125 V, Three wire, grounded, two pole

20 A, 250 V, Three wire, grounded, two pole

15 A, 250 V, Three wire, grounded, two pole

15 A, 125 V, Three wire, grounded, two pole

15 A, 250 V, Three wire, grounded, two pole

15 A, 125 V, Three wire, grounded, two pole

9-23. A single and duplex receptacle for various current ratings and voltages. (Thomas & Betts)

9-24. A switch box for Romex. It can be mounted without nails. The bracket on the side has an "eagle claw" so that the tabs may be hit with a hammer and driven into the stud without using nails. Note the beveled corner and clamps for the Romex. (Appleton)

9-25. A square-corner Romex switch box, with clamps. This one is deeper than the standard size. (Appleton)

9-26. This beveled-corner switch box for Romex may be mounted without nails. (Appleton)

9-27. A square-corner switch box with Romex clamps inside. This one does not have plaster ears, but has side leveling ridges and tapped grounding holes. (Appleton)

9-28. A bevel-corner Romex switch box, with tapped grounding holes and side leveling ridges. Nail holes are located on the brackets outside the box. A Romex clamp is included. (Appleton)

9-29. *Two switch boxes, ganged. They have Romex clamps and claws, and nails. The claws may be hit with a hammer to hold the box in place until nails can be inserted. (Appleton)*

9-30. *Note how the bare bonding wire is attached to the green hex grounding screw on the receptacle. This view also shows the location of clamps inside the box to hold both the BX and the Romex. (Harvey Hubbell)*

9-31. *A square-corner Romex switch box without plaster ears, but with side leveling ridges and the side nail capability. (Appleton)*

• *A handy box or utility box.* These are designed to be used with either thinwall or rigid conduit. They can handle a switch or an outlet. If the knockout is removed and the Romex cable is properly protected, a box may be used on exposed surfaces if a connector to hold the Romex inside the box is added. Fig. 9-36.

Figure 9-37 shows the following procedures: *A* shows how to connect the existing third wire (green insulated or bare copper) to the green grounding screw on a receptacle; *B* shows how the ground wire from the green grounding screw should be connected to a water pipe if no ground is available in the existing system; *C* shows that the outlet is automatically grounded if properly grounded armored cable (BX or EMT) is used throughout the system. The grounded outlet provides protection against shock hazards, such as defective internal wiring on a hand drill.

9-32. A switch box with nailing holes and hammer driving the nails into the stud. (Appleton)

9-33. A beveled-corner switch box mounted with two nails. (Appleton)

9-34. A front view of a mounted beveled-corner switch box. (Appleton)

9-36. *A handy box (utility box) with or without nailing brackets for thinwall or rigid conduit. Galvanized. (Appleton).*

9-35. *An insulated wire grounded to the switch box by using the tapped grounding screw hole provided. The insert shows the wire attached to the box by a screw. (Appleton)*

TO GROUND

NONMETALLIC CABLE
(CAN ALSO BE USED WITH
METALLIC CABLE)

C

BOX GROUND
SCREW

GROUNDING
STRAP

BONDING JUMPER
ELIMINATED

NOTE: NEC ARTICLE 210-7(a) EXCEPTION NO. 1.
GROUNDING TYPE RECEPTACLES OF THE TYPE
THAT REJECT NONGROUNDING TYPE ATTACH-
MENT PLUGS OR WHICH ARE OF THE LOCKING
TYPE SHALL BE PERMITTED FOR SPECIFIC
PURPOSES OR IN SPECIAL LOCATIONS.

9-37. *Grounding of a receptacle. (Circle F)*

GREEN
GROUNDING
SCREW

SPRING-TYPE GROUNDING
STRAP FOR HOLDING
MOUNTING SCREW

A

WATER PIPE HAS TO HAVE
A POSITIVE GROUND

B

WATER PIPE

ARMORED
CABLE

THIRD
WIRE

NOTE: ARTICLE 210-7(d) EXCEPTION: WHERE
A GROUNDING MEANS DOES NOT EXIST IN
THE RECEPTACLE ENCLOSURE A NON-
GROUNDING TYPE OF RECEPTACLE SHALL
BE USED.

NOTE: A RECEPTACLE DESIGNED WITH
A SPRING-TYPE GROUNDING STRAP
WHICH HOLDS THE MOUNTING SCREW
CAPTIVE AND ELIMINATES A BONDING
JUMPER TO THE BOX IN ACCORDANCE
WITH EXCEPTION NO. 2 OF SECTION
250-74 DOES NOT REQUIRE THE
GROUNDING WIRE TO BE CONNECTED
TO THE GREEN GROUNDING SCREW.

Romex is protected in an exposed position against a wall in a basement. Fig. 9-38.

• *A handy-box extension.* Sometimes this is added if the proper volume is not available for the necessary number of conductors making their entrance and exit from the utility box. An extension simply slips onto the screws of the utility box and is tightened into place. A cover plate or switch is then placed in the box, or over the extension. Fig. 9-39.

• *Cover plates.* Because utility boxes must be totally enclosed, several plates are available to fit individual situations. Figures 9-40 through 9-43 show the types available.

9-41. A handy box cover with three knockouts. Captive screws and a galvanized finish make up the total cover. "Captive screws" means they will not fall out when unscrewed from a handy box. They must be intentionally unscrewed to remove them from the cover. (Appleton)

9-39. A handy-box extension, needed where the box capacity is not sufficient to allow the number of wires or conductors required. This extension slips onto the two screws on the handy (utility) box. It is an exact duplicate of the handy box, but without a bottom. (Appleton)

9-38. A handy box being used with ½-inch thinwall to protect the Romex from damage. This type of installation is usually found in a home in the basement where there is the possibility of damage to the unprotected Romex by the movement of heavy objects.

9-42. A handy-box switch box cover for tumbler or toggle switch with square handles. (Appleton)

9-40. A blank cover for the handy box. (Appleton)

9-43. A handy box cover for a 20-ampere, single-receptacle outlet. (Appleton)

9-45. A 4-inch octagonal box for Romex or BX. Note the tapped grounding screw holes. Clamps are included. This type is not to be used in ceilings without mud rings. (Appleton)

9-46. A 4-inch octagonal outlet box for Romex and BX with bracket. No nails are needed for mounting. This one includes clamps and has a tapped grounding screw. (Appleton)

9-44. How to install a switch plate. (Circle F)

Figure 9-44 gives directions for installing a switch plate on a box. Make sure the power is turned off if you are installing a plate on existing wiring. Note the different types of plates: double switch plate, combination switch plate and double-outlet plate, double-outlet plate, and switch plate and single-outlet plate.

Octagonal boxes of 3- and 4-inch sizes are available for use with Romex, BX, thin-wall, or rigid conduit. They come with or without clamps and have knockouts of various sizes. All of them have two screws for attaching a cover plate. Extensions are available for increasing the space to hold a number of conductors. Figs. 9-45 through 9-49.

Sometimes it is necessary to place a utility box or outlet box between two studs or joists. For this situation, an expandable bar hanger comes in handy. It expands to fit

9-47. An outlet box used as a junction box for Romex and flexible metal conduit. The cover is blank.

9-48. A fixture hanger with safety clips. (Appleton)

9-49. *A 3-inch octa-gonal-box extension ring. (Appleton)*

9-50. *An expandable bar hanger. It can be adjusted for 14-, 16-, 18- or 20-inch centers. (Appleton)*

9-51. *A bar hanger, 18 inches long, with slid-ing clip. (Appleton)*

9-52. *An outlet box mounted on an expand-able bar hanger. (Appleton)*

between studs or joists centered 14, 16, 18, and 20 inches. The center clip is adjust-able to any location between the maximum and minimum setting. Fig. 9-50. The bar hanger in Fig. 9-51 has nail holes for mounting between studs or joists. The sliding clip facilitates easy location of the box. Figure 9-52 shows a Romex bar set with an outlet box. It will expand to fit joists on 14-, 16-, 18-, or 20-inch centers.

Occasionally, a cover is needed that will extend a box to fit flush with the wall. A raised cover ($\frac{5}{8}$ inch) for a 4-inch octagonal outlet box is shown in Fig. 9-53. It is also available in other sizes.

Sometimes an outlet with a steel cover plate is required for areas subject to possi-ble mechanical damage of the plate. A steel cover plate for a 4-inch octagonal box will accommodate a duplex receptacle. Fig. 9-54. Figure 9-55 shows a blank cover plate for a 4-inch octagonal box. However, a blank box cover is not always the answer, espe-cially if you want to mount an object from the box. Thus, a blank cover with a $\frac{1}{2}$-inch knockout might be utilized, so that a fixture may be mounted from the 4-inch octago-nal box it covers. Fig. 9-56.

9-53. *A 4-inch octagonal outlet box with a $\frac{5}{8}$-inch raised cover. (Appleton)*

9-54. *A 4-inch octagonal outlet box cover with flush mounting of a duplex recept-able. (Appleton)*

9-55. *A 4-inch octagonal outlet box cover, blank. (Ap-pleton)*

9-56. A 4-inch octagonal outlet box cover with a ½-inch knockout. (Appleton)

9-57. A 4-inch square drawn box with tapped grounding screw holes and two screws for cover mounting. This one is made for use on rigid conduit or thinwall conduit, but can also be used for Romex. (Appleton)

9-58. A 4-inch square box for thinwall or rigid conduit. It can also be used for Romex if the proper connectors are used. (Appleton)

9-59. This 4-inch square box has a bracket for mounting against the stud without nails. (Appleton)

9-60. A 4-inch square box for thinwall, Romex, BX, or rigid conduit. (Appleton)

9-61. A 4-inch square box extension. Note how it fits over an existing box. (Appleton)

In Figs. 9-57 through 9-61 you will find an assortment of boxes 4-inch square, some with tapped, grounding-screw holes and some with clamps. Some of these boxes have no brackets for mounting; others have no-nail brackets and self-alignment brackets. Some use nails for mounting. Others will be used on conduit or thinwall by removing the knockout and inserting the proper connector. There is also an extension for these boxes. A box that is $4\frac{11}{64}$-inch square has a greater cubic-inch volume for additional conductors. It is difficult at first to recognize the difference that only $\frac{11}{64}$ inch makes.

Once you are working with both the 4-inch and the $4\frac{11}{64}$-inch sizes, however, you will become very accustomed to the difference.

Covers for a 4-inch square box are many and varied. Figures 9-62 through 9-66 show various types and configurations. Note figure captions to obtain more details of size, function, and shape.

FIXTURES AND SPLICES

Lighting Fixtures

Lighting fixtures may be installed in several ways. There are hanger supports that thread onto a threaded stud, mounted in the box, as in Fig. 9-67. Straps and machine screws are used to mount the fixture in Fig. 9-68. If there is no stud, the metal strap may be used to hold the canopy of the light fixture in place, as in Fig. 9-69.

Glass-enclosed ceiling fixtures are easily attached to the ceiling box by using a threaded stud. Fig. 9-70. Wall lights may be installed using the same method. Make

9-62. A $\frac{1}{2}$-inch deep surface cover for a 4-inch square box. This one will accommodate one duplex flush receptacle. (Appleton)

9-63. A surface cover with toggle switch and duplex holes. (Appleton)

9-64. A switch cover for a 4-inch square box. (Appleton)

9-65. A $\frac{1}{4}$-inch raised single-device cover for a 4-inch square box. It is used where a larger box is needed to hold the wires, but a single receptacle or switch is mounted in the box. (Appleton)

9-66. A flat cover blank for use on a 4-inch square box. (Appleton)

9-67. Mount large drop fixtures by simply using a screw hanger support onto the threaded stud in the outlet box. Use solderless connectors (wire nuts) to connect the electrical wires and a grounding clip for the extra uninsulated ground wire. Raise the canopy and anchor in position by means of a locknut. (Sears)

9-68. Outlet box has a stud in this case. Insert the machine screws in threaded holes of the metal strap shown. Slip the center hole of the strap over the stud in the outlet box. Hold the strap in position by a locknut. Connect wires with wire nuts and slip the canopy over the machine screws; fit flush and secure the fixture with two cap nuts. Don't forget to anchor the uninsulated ground wire to the box with a ground clip or to the box's threaded grounding screw. (Sears)

9-69. If there is no stud, insert the machine screws as shown here. Fasten the ears of the outlet box and the strap with screws. Then align the canopy onto the two screws pointing down and cap off with cap screws. (Sears)

9-70. Glass-enclosed fixtures can be installed by the method shown here. (Sears)

9-71. Wall brackets or lights are installed by strapping to the ears of the box, then using a nipple and cap to complete. Don't forget the ground wire and the ground clip. If there is an outlet which is "always on," wire according to the insert. (Sears)

9-72. *Replacing a damaged lampcord plug with an attachment plug. Note the underwriter's knot and how it is looped to prevent strain on the connection when the cord is pulled.*
(Circle F)

sure the outlet found in the wall fixture is wired for full-time service — not controlled by the switch that turns the light on and off. Fig. 9-71.

Replacing a Damaged Lampcord

Plugs and cords sometimes need repair. Replacement plugs with a heavy-duty capability should be used. Figure 9-72 shows how the *underwriter's knot* is used to relieve strain put on the connection when the cord is pulled. The two plugs are replacement types that do not have a third wire for grounding.

Some lamps use an *in-line switch* for control. It may be replaced easily, or installed, if needed, by following the steps shown in Fig. 9-73.

9-73. *Some lamps have an in-line switch for controlling the light bulb. A switch can be put in the line easily, as shown. Make sure the power is off or the plug has been pulled before attempting to work on the line.*

An adapter is available for three-pronged plugs to make them usable on the older two-wire systems installed in most homes before 1960. Fig. 9-74. The lug is inserted under the screw that holds the cover plate onto the outlet.

Splicing Wires

Manufacturers of a specific type of fastener or clip have specialized methods that they recommend to splice wire. Most of these will be covered in Figs. 9-75 through 9-93 (Fig. 9-93 on Page 216). Read the captions under the figures to learn the proper procedure for making and insulating a splice.

9-74. This three-wire grounding adapter makes it possible to use three-pronged safety grounded plugs with a conventional two-wire outlet. Make sure the green insulated grounding wire attaches to the outlet box by being placed under the screw which holds the cover plate in place. Refer to Figs. 4-21 and 4-22. (Circle F)

9-75. To splice wires: 1. Insert untwisted stripped wires through the splice cap. 2. Twist wires. 3. Cut wires flush with cap. 4. Insert tool—squeeze to crimp. 5. Snap on a nylon insulator. (Buchanan)

9-76. Splicing solid and stranded wires: 1. Loop stranded wire, if No. 16 or smaller. 2. Loop solid wire, if smaller than stranded wire. 3. When joining two or more solid wires to a larger stranded wire, twist the solid wires together. (Buchanan)

9-77. To insulate splices, just snap on the nylon insulator over the installed splice cap. (Buchanan)

RING TONGUE

SPADE TONGUE

LOCKING TONGUE

9-78. *To terminate wires with lugs: 1. Strip wire(s) approximately $\frac{5}{16}$ inch. 2. Slip lug on wire with tool in position. 3. Insert in tool with flat side of the lug in the "up" position so that the tongue enters the slot in the latch. 4. Squeeze the tool to crimp. (Buchanan)*

9-79. *To remove splicing caps or lugs without damaging the wires, snip splice cap at both ends of one crimp, counteracting cutting pressure with the index finger. Then peel off the cap. To remove the lugs, snip off tongue and proceed as for the splice caps. (Buchanan)*

9-80. *To remove "cold working" cap, apply pressure alternately at two points 90° apart and between crimps — then pull off the cap. This is most effective when the cap is not full of wire. (Buchanan)*

9-81. *To remove nylon insulators, apply pressure alternately at two points 90° apart to "cold work" metallic retainer — or cut up side of insulator body. (Buchanan)*

9-82. *To terminate No. 6 wire using two lugs: 1. Strip wire approximately $\frac{3}{8}$ inch. 2. Untwist wire lay and separate strands into two approximately equal groups. 3. Crimp a lug on each group. 4. Bring flat sides of lugs together. (Buchanan)*

9-83. *How to make a strain-relief splice for service entrance: 1. Strip both wires approximately $1\frac{1}{2}$ inches and untwist wire lay; install two splice caps leaving about $\frac{1}{16}$ inch between the two caps. 2. Cut off the top half of the tip of an insulator and push remainder over the cap. 3. Snap another insulator over the end cap. (Buchanan)*

213

9-84. Tee-tap (where slack permits): 1. Strip and loop wire. 2. Add stripped tap wire, splice, and snap on the insulator. (Buchanan)

9-85. Adding wire (No. 12 or smaller): 1. Loop wire to be added and press into the crimped indentation of the cap. 2. Force the splice cap over the joint, then crimp. 3. Insulate with a snap-on insulator. (Buchanan)

9-86. Grounding connection. Splice wires with a splice cap, leaving one wire extending through the cap to permit attachment of a lug. (Buchanan)

9-87. Terminating two or more wires with a lug. This is a technique used if the barrel capacity of the lug is too small to hold the wires. Splice wires, leaving one or more wires extending through the cap to permit attachment of the lug as shown. (Buchanan)

9-88. Wire nut insulator used for a splice of wires inside a switch box or outlet box. (3M)

9-89. Wire nut insulators used in various sizes for different wire sizes. They are also available in color for coding of wires. (Ideal)

9-90. Wire nuts with a Bakelite case and copper inside. (Ideal)

9-91. Using wire nuts for the splicing needed in the installation of fluorescent fixtures. (3M)

9-92. Two- and three-wire splices with plastic "see-through" insulators. (Buchanan)

215

9-93. *Various sizes and shapes of end caps for splicing wires. (Buchanan)*

Mounting Boxes in Existing Walls

A switch box may be mounted by using the existing wall as part of the support device. Figs. 9-94 and 9-95. A special clip that will exert pressure from the inside of the wall when the screws are tightened is shown in Fig. 9-94. In Fig. 9-95 a piece of sheet metal is inserted to fold over inside the box and fit behind the dry wall to support the box. Note that the plaster ears on the box are important. They add to the support when a switch is turned on and off or when a plug is pulled from an outlet mounted in the box. These are two of the many methods used to add an outlet to an existing wiring system.

NUMBER OF CONDUCTORS IN A BOX

Every box used for switches, receptacles, or fixtures has a volume. This volume determines the number of wires that can be brought into the box safely. The National Electrical Code has a table of sizes and the maximum number of wires (conductors)

9-94. *Using a box with "grip-tight" brackets to hold a box in place in the addition of a switch or outlet to existing wiring. (Sears)*

9-95. *Metal box supports can assure a stronger job. Insert supports on each side of the box. Work supports up and down until they fit firmly against inside surface of the wall. Bend the projecting ears so that they fit around the box. (Sears)*

Table 9-A. Boxes and Maximum Number of Wires*

Box dimension, inches trade size or type	Min. cu. in. cap.	Maximum number of conductors			
		#14	#12	#10	#8
4 × 1¼ Round or Octagonal	12.5	6	5	5	4
4 × 1½ Round or Octagonal	15.5	7	6	6	5
4 × 2⅛ Round or Octagonal	21.5	10	9	8	7
4 × 1¼ Square	18.0	9	8	7	6
4 × 1½ Square	21.0	10	9	8	7
4 × 2⅛ Square	30.3	15	13	12	10
4¹¹⁄₁₆ × 1¼ Square	25.5	12	11	10	8
4¹¹⁄₁₆ × 1½ Square	29.5	14	13	11	9
4¹¹⁄₁₆ × 2⅛ Square	42.0	21	18	16	14
3 × 2 × 1½ Device	7.5	3	3	3	2
3 × 2 × 2 Device	10.0	5	4	4	3
3 × 2 × 2¼ Device	10.5	5	4	4	3
3 × 2 × 2½ Device	12.5	6	5	5	4
3 × 2 × 2¾ Device	14.0	7	6	5	4
3 × 2 × 3½ Device	18.0	9	8	7	6
4 × 2⅛ × 1½ Device	10.3	5	4	4	3
4 × 2⅛ × 1⅞ Device	13.0	6	5	5	4
4 × 2⅛ × 2⅛ Device	14.5	7	6	5	4
3¾ × 2 × 2½ Masonry Box/Gang	14.0	7	6	5	4
3¾ × 2 × 3½ Masonry Box/Gang	21.0	10	9	8	7
**FS—Minimum Internal Depth 1¾ Single Cover/Gang	13.5	6	6	5	4
**FD—Minimum Internal Depth 2⅜ Single Cover/Gang	18.0	9	8	7	6
**FS—Minimum Internal Depth 1¾ Multiple Cover/Gang	18.0	9	8	7	6
**FD—Minimum Internal Depth 2⅜ Multiple Cover/Gang	24.0	12	10	9	8

*Reprinted by permission from NFPA 70-1981, *National Electrical Code,* © 1980 National Fire Protection Association, Boston, Massachusetts
**FS and FD are designations used for Unilets with conduit.

allowed in a box. This is one of the most important things you must consider when you are wiring a house. The inspector looks here to make sure you have done the job properly.

Table 9-A shows the number of wires in a box. This assumes all the wires are the *same* size. If the wires are not the same size, you will have to check the volume of the box and then take a look at Table 9-B. Here, the volume required for each wire is shown. Just add the number of wires of a particular size and multiply by the volume

Table 9-B. Volume Required for Each Wire Size*

Size of conductor	Free space within box for each conductor
No. 14	2.00 cubic inches
No. 12	2.25 cubic inches
No. 10	2.50 cubic inches
No. 8	3.00 cubic inches
No. 6	5.00 cubic inches

*Reprinted by permission from NFPA 70-1981, *National Electrical Code,* © 1980 National Fire Protection Association, Boston, Massachusetts

Max. 8 in.

Staple

Min. 1/4 in.

9-96. *Installation of the nonmetallic (NM) wire known as Romex in a single-gang box.*

required by that conductor. Then do the same thing for the next size conductor. Once you have figured all the various sizes and their volume requirements, just add them. This will tell you whether or not you have too much for the volume of the box.

Examples: A 4- × 1¼-inch round or octagonal box will have a minimum cubic inch capacity of 12.5. If you use all No. 14 wires, it can handle six of them. If you use No. 12 wires, it can hold only five wires. However, if you mix the two sizes, you have to go to Table 9-B to see what the volume requirement of each conductor is. If you look at the table you will see that the No. 14 wire must have 2.0 cubic inches for each conductor. The No. 12 wire will need 2.25 cubic inches for each conductor. You have only 12.5 cubic inches in the box. Therefore, you can have only three wires of No. 12 for 6.75 cubic inches and two wires of No. 14 at 4 cubic inches total. Add them up and you get 10.75 cubic inches. If you use three of No. 14 and three of No. 12, you get 12.75 cubic inches, or 0.25 cubic inches more than the box has to offer. Therefore, the three No. 12 and three No. 14 wires are not permitted in this size box. You could go to the next size box, which is a 4- × 1½-inch round or octagonal. Such a box has 15.5 cubic inches capacity.

Which Wires to Count

● *Ground wires.* No matter how many ground wires come into a box, a deduction of only one conductor must be made from the number of wires shown in Table 9-A. Any wire that runs unbroken through a box is counted as one wire. The ground wires may be in nonmetallic cable or ground wires that are run in metal or nonmetallic raceways.

● *Spliced wires.* Any wire that comes into a box and is spliced is counted as *one* wire. It may be crimped or twisted in its connection. If a wire that comes into the box is connected to a wiring device terminal, it is counted as *one* wire.

● *Clamps in boxes.* Cable clamps, hickeys, and fixture studs count as one wire whether the box has one clamp, two clamps, or any combination of clamps, studs, or hickeys.

You may remove unused cable clamps from a box to provide more space in the box. If one clamp is left in the box, you must count the clamp as one conductor. If both clamps are removed and you use a box connector, then the clamp is not deducted from the total allowed in the box.

● *Grounding jumper.* If a jumper is used from the box screw to the receptacle grounding terminal, then the jumper is not counted as a conductor since it does not leave the box.

If a switch or receptacle has a grounding strip on it, this must also be counted as a conductor in terms of the number of conductors used in a box.

● *Mounting boxes.* One of the factors involved in mounting boxes and electrical equipment is the prevention of fire. Also, in case of fire, it must not be allowed to move from one level to another by way of a raceway or hole. The use of a hexagonal or square box in ceilings where there is sheetrock is not permitted. If they are used, they must have a "mud ring" installed to prevent fire from being allowed to contact wall or ceiling materials that may burn.

● *Single-gang boxes.* In single-gang boxes, the nonmetallic cable does not have to be clamped to the box. It should be secured to a stud within 8 inches of the box, however. Fig. 9-96.

If the box is round, or square-two-gang or three-gang, the cable must be clamped to the box. Fig. 9-97.

Square   Two-Gang

9-97. *Boxes that must have a clamp or connector if used for wiring.*

Round

Three-Gang

● *Installing a connector.* Figure 9-98 shows the newer type of Romex box connector. There are no screws needed. Insert the Romex cable as shown in Fig. 9-99. Slide the connector over the cable from inside the box. Then, squeeze the connector on the outside of the box to make a fit with no movement of the wire or connector.

ALUMINUM CONNECTIONS

Aluminum is a more active metal than copper. In the presence of moisture, the aluminum will erode. This susceptibility to galvanic corrosion, or electrolysis, is a problem because it will weaken an aluminum connection. Fig. 9-100.

The electrolytic action between aluminum and copper can be controlled by plating the aluminum with a neutral metal (usually tin). Such plating prevents electrolysis from taking place. It also helps keep the joint tight. As an additional precaution, however, a joint-sealing compound should be used. The compound contains fine zinc particles which break through the oxide film that forms on an aluminum connector. It also seals out air and moisture.

Upon exposure to air, aluminum immediately becomes coated with a film of oxide. This oxide film has the properties of a ceramic, therefore insulating the aluminum and increasing joint resistance. The film must be removed, or penetrated, before a reliable aluminum joint can be made.

Aluminum connectors are designed to bite through the film of oxide on the aluminum as the connector is applied to the conductors. It is further recommended that the conductor be wire-brushed and preferably coated with a joint compound to guarantee a reliable joint.

9-98. *Romex connectors — no screws, no lock-nut. Just use a pair of pliers to assert pressure to hold the Romex in place. (Buchanan)*

insert cable
slide connector
over cable

squeeze

9-99. *Insert the Romex cable; slide the connector over the cable from inside the box. Then squeeze the connector on the outside of the box to make a tight fit with no movement of the wire or connector. (Buchanan)*

9-100. *Electrode potential. (Values show only the relative activity of metals. Activity decreases in going from copper to aluminum, tin-plated copper to aluminum, copper to tin-plated aluminum, tin-plated copper to tin-plated aluminum.) Cu = copper, Al = aluminum, Sn = tin. (Thomas & Betts)*

Aluminum Wire for Home Wiring

Aluminum wiring came into use when there was a shortage of copper wire. Many houses built between 1965 and 1971 used such wiring. The aluminum wire was usually No. 12 instead of the copper No. 14.

Use of aluminum wire can be a safety hazard. The greatest problems have occurred where there are extreme temperature differences. Aluminum expands and contracts at a greater rate than copper. Most of the potential problems can be eliminated by tightening all connections at outlets and switches. Materials now available can ensure that copper and aluminum can be safely connected with wire nuts or by other means.

There are two types of aluminum wire in general use today: all-aluminum and copper-clad aluminum. Bonded to the surface of the copper-clad aluminum is a coating of copper. This eliminates some of the problems that occurred when copper and aluminum were placed together in a wire nut or some other connector.

When aluminum wire was first introduced, it was used just as copper wire. This created some problems, since the aluminum wire was clamped in a brass or copper terminal screw or connector. The heat created by current flowing in the wire causes the aluminum to *cold-flow* out of the connector. When the connection cools, it becomes a little loose. As the current is turned on and off, the heating and cooling cause still more aluminum to become extruded. Eventually the wire can heat up and cause a fire.

In damp areas this can cause an electrolytic action to take place. Copper or brass terminal screws react with the aluminum.

All switches and receptacles that were intended for 15- or 20-ampere service are not marked. However, those intended for use with copper (CU) and aluminum (AL) are marked CU-AL. Recently the UL changed their rating of 15- and 20-ampere switches and receptacles and introduced the CU-ALR designation for aluminum in the 15- and 20-ampere rating. The *R* stands for *revised*. The composition of the aluminum wire has been changed also. There are special chemical and physical properties specified by the UL before the wire gets approval.

Devices marked CU-AL should be used with copper wire only.

Devices marked with CU-ALR (or CO-ALR) may be installed with aluminum, copper-clad aluminum, or copper wire. Devices with a push-in connection must *not* be used with all-aluminum. They can be used on copper-clad aluminum or copper wire.

Devices rated 30 amperes or more are *not* marked with CU-ALR or CO-ALR. Those not marked, and rated at 30 amperes or more, can be used with copper-clad aluminum or copper wire.

CIRCUIT PROTECTIVE DEVICES

The circuit breaker and the fuse are the circuit protective devices most frequently used to ensure that a circuit operates within its designed limits.

Fuse

The fuse is a device that destroys itself when it operates. That is, a fusible link is destroyed when it performs its designed function. An overcurrent will cause a piece of metal in the fuse to heat up and finally melt, thereby interrupting the circuit. This occurs because the fuse is inserted in series with the device it is to control.

One type of fuse is shown in Fig. 9-101. It is a spring-loaded one that can disrupt the circuit if a unit overheats and is removed from its contact, as in (d), or if it burns or melts, as in (c). Part (a) of Fig. 9-101 shows the unit in its complete form, with a screw-in base for insertion in a fuse box. A cutaway view with the construction of this type of fuse is shown in (b). These are normally found in outmoded wiring in old homes.

Circuit Breakers

A circuit breaker is a device which, after breaking a circuit, can then be reset by turning the handle to *off* and then to *on*. Fig. 9-102.

(a)　　　(b)　　　**9-101.** *A screw-in fuse and its operation.*　　　(c)　　　(d)

The circuit breaker is mounted in a distribution box by snapping it into place. The hot wire (black or red) is attached by inserting it under the screw and tightening. A knockout blank in the distribution box must be removed, to allow for the handle and top of the circuit breaker to be exposed. Figure 9-103 shows a circuit breaker for 20 amperes and a single connection.

Another type is the 100-ampere double circuit breaker, with the two breakers tied together, or a common trip for a two-pole arrangement. Fig. 9-104. It requires two spaces in a distribution box.

Circuit-breaker box. The circuit breaker must be housed to protect it from abuse. It also has exposed areas that could cause shock if a person touched them when the circuit was live. The opening for the cable to be fed into the box is easy to accomplish as

9-102. *A cutaway view of a circuit breaker. (Bryant)*

9-103. *A single 20-ampere circuit breaker. (Bryant)*

9-104. *A double-circuit breaker, 100-ampere, with common trip. (Bryant)*

223

9-105. *Removing the knockout from a panel box for circuit breakers. (Square D)*

A   B   C

shown in Fig. 9-105. The knockout is removed by using a screwdriver and a pair of pliers. Then the connector of the proper size is used to hold or clamp the wire so it will not be moved easily by an external force.

Placing the circuit breakers into the panel must be done by removing the slots that allow the circuit breaker to protrude through the box for easy operation. By placing this panel over the circuit breaker, it is possible to protect the operator of the circuit breaker from exposure to 240 volts within the circuit-breaker box. Fig. 9-106.

Tap Screwdriver Here.

Then Tap Screwdriver Gently Here

9-106. *Using a screwdriver to take out the slots in a front panel of a circuit-breaker box. Note how the screwdriver is first tapped on the end at one position and then the other. This knocks the slot material loose and it can be taken out by hand. (Square D)*

QUESTIONS

1. What size Romex is used in house wiring?
2. Why is box volume important?
3. What is BX cable? How is it different from Romex?
4. What is a utility box?
5. How do you replace a lamp cord with an in-line switch?
6. What devices do you use to mount boxes in existing walls?
7. When counting wires in a box, how many times do you have to count the ground wires that come into the box from different directions?
8. What is meant by a spliced wire?
9. What is a cable clamp? Where is it used?
10. Why are square and hexagonal boxes not allowed in ceilings where there is sheetrock?
11. What is a Romex connector?
12. What causes an aluminum wire connection to weaken?
13. Why must aluminum wire have a larger diameter than copper wire to carry the same amount of current?
14. What does CU-ALR on the insulation sleeving of wire mean?
15. What is the advantage of a circuit breaker over a fuse?
16. What is another name for an over-current device?

Low-Voltage and Lighting Circuits

Other chapters have discussed various "standard" methods of wiring a house or building. Here we will discuss low-voltage remote control for home use.

LOW-VOLTAGE REMOTE CONTROL

Low voltage is used in some applications where industrial plants, homes, or offices require remote control of a piece of equipment or a light. Usually a 24-volt source is used to control relays, which in turn, control the 120-volt or 240-volt circuits.

A *relay* is a device that has a coil of wire and contacts. The coil of wire has a loosely fitted core. This core can move when the coil is energized. The coil has a *sucking effect* that tends to pull the core into the coil. If a pair of contacts is attached to the core, the contacts can control the other circuit. These contacts can be controlled by energizing or deenergizing the coil. Low-voltage coils are used with small wires and switches. With low voltage it is possible to use small wires and a different type of wiring.

Control Flexibility

With the advent of the three-way switch, the concept of multi-point switching gradually broadened into the many conveniences and economies of modern automated lighting controls. A single switch at more than one point can control a single lighting fixture. Several switches at one point can control many lighting fixtures at various locations. These may be either on the same or on different branch circuits.

Where desirable, a master switch can start a motorized control unit. It turns any number of lights *on* or *off* as required. The addition of a program timer allows several motor-master control units to be set into motion at one time. It will turn on selected lights, then turn off all lights left on. At the end of the day, photoelectric controls can override these program timers. This occurs when light intensities fall below preset levels. The flexibility provided by low-voltage, remote-control switching is practically limitless.

Use in Modern Construction

Relay-switching circuits have low-voltage, low-current requirements. Thus, small, flexible wires may be snaked in thin wall or steel partitions without the protection of metal raceways (except where local codes prohibit such installations). Wires can be dropped through hung ceilings into movable partitions as easily as rewiring telephones. In fact, wiring can be run under rugs to switches located on desks, and easily changed when required. The switches are modern in appearance and very compact.

The low-voltage systems in this chapter are those designed by General Electric Company. However, there are other manufacturers, and the systems are basically the same. The products here have been chosen to give a relatively typical approach to the general idea of using low voltage for remote-control circuits.

SYSTEM COMPONENTS. There are a number of component parts familiar to the electrician. These low-voltage devices have been redesigned to accommodate particular situations that can easily utilize the convenience and reliability of such systems.

• *Power supply.* All the components of the GE system operate on 24 volts. To obtain this voltage, it is necessary to step down the usual 120 volts available in homes. For this purpose, special transformers with energy-limiting characteristics are available for either 120 VAC (volts AC) or 277 VAC operation. Fig. 10-1.

10-1. *A transformer, available in 115 VAC or 277 VAC. It steps down to 24 volts. (General Electric)*

The transformer has two pigtail leads that can be spliced to a 120-volt line with wire nuts. Note that the cover-mounted transformer has to be mounted on the outside of the utility box. This means that the low voltage will be outside the box, and the high voltage (120 volts) will be inside, according to Code. Low- and high-voltage lines should not be included in the same box or conduit.

• *Relays.* Low-voltage switching depends upon relays to do the work. The low-voltage switching system eliminates line-voltage wiring to all switches and replaces it with inexpensive, low-voltage wire. The heavy work of switching is done by dependable relays that can be controlled from various remote points. Master switches control many circuits from one, or from several convenient locations. Fig. 10-2.

Relays used here are of the mechanical latching type. Thus, the switching circuits require only a momentary impulse—as short as a half-cycle or $\frac{1}{120}$th of a second. These relays have a coil design that resists burnout due to equipment or operational failure. This usually results in energizing the relay for extended periods at ambient temperatures not exceeding 60°C [140°F.]. They can control 20 amperes of tungsten, fluorescent, or inductive loads at 125 VAC and 277 VAC. They can also be used for $1\frac{1}{2}$ HP 240 VAC and $\frac{1}{2}$ HP 125 VAC. Some relays are available with a pilot light switch to indicate the *on* position in remote locations on a master panel.

125 Volt AC (alt. current)

10-2. *The remote control wiring system eliminates wiring to all switches and replaces it with inexpensive, low-voltage wire. (General Electric)*

Figure 10-3 shows the internal wiring of a relay. The enclosed relay, shown in Fig. 10-4, may be mounted in a metal box by removing a 1-inch knockout and sliding the relay through the hole. It will snap in place for permanent mounting.

• *Switches.* One basic type of switch, the single-pole, double-throw (SPDT) momentary-contact, normally open circuit, does the work of single-pole, double-pole, three-way or four-way switches used in conventional wiring. It also uses much less wiring and of a smaller size. The switches are in the circuit for a split-second, just enough to energize the relay and have it move the contacts to the *on* or *off* position. Many switches may be wired in parallel to control either a motor or a light from a number of locations. Switches are available with or without a pilot light.

Figure 10-5 is a pilot light switch with push-button control action. Figure 10-6 is an interchangeable line switch. It fits a standard or a narrow interchangeable mounting bracket. The center position does not make contact. Up is the *on* position; down, the *off* position. A heavy-duty switch with the possibility of fitting into a standard wall-box mounting is shown in Fig. 10-7. Here, again, the center position is not making contact.

10-3. *The internal wiring of a relay. (General Electric)*

BLUE
BLACK
YELLOW
RED

10-5. A single-pole, double-throw, momentary-contact, normally open circuit switch. (GE)

10-4. A relay for low-voltage control circuits. (GE)

10-6. This remote-control switch matches other kinds of interchangeable switches and fits standard or narrow mounting brackets and wall plates. It is normally open, single-pole, double-throw, momentary-contact. (GE)

10-7. A heavy-duty switch for standard wall-box mounting. It is available with a rating of up to 30 amperes. (GE)

10-8. *A simple remote-control circuit. (GE)*

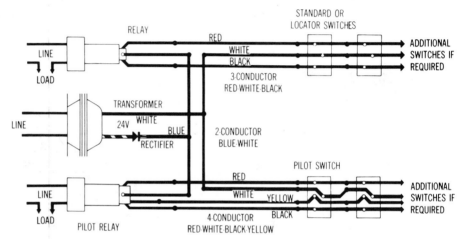

10-9. *The basic circuitry for a remote-control system. (GE)*

When pushed momentarily into the up position, it is *on;* down is *off.* The contacts can handle up to 30 amperes.

• *Circuits.* Figure 10-8 shows the simple utilization of one switch, one transformer, and one relay. The circuit is capable of switching a 120- or 277-volt circuit without an operator being near. This means that high-voltage switching (277) may be done outside the area, not in the control center where the operator contacts the switch for controlling action. The safety of such an arrangement is evident.

The red, white, and black wire from the switch to the relay and transformer is usually small—about No. 20 AWG (American Wire Gage). This means it can be run longer distances for less cost than conventional wiring.

Two relays sometimes control two different circuits. Fig. 10-9. Note that a *rectifier* is used to change the 24 volts AC to DC, for less noise in the operation of the relays. A rectifier is a device that changes AC to DC.

• *Rectifiers.* A semiconductor diode is used as the rectifier to change the 24-volt, AC transformer output to DC. Note that it is connected to a terminal strip, for ease in mounting as well as making connections. This rectifier is capable of 7.5 amperes continuous duty, and 20 amperes for short periods. Fig. 10-10.

• *Remote-control wire.* Figure 10-11 shows a three-conductor insulated wire for low-voltage, remote-control wiring. The seven strands of copper wire make up a No. 20 AWG conductor having $\frac{1}{64}$ inch of thermoplastic insulation. The three-conductor wire has a red stripe on one outer conductor and a black stripe on the other outer conductor.

In old construction, wire can be snaked into walls, or stapled to wallboard surfaces. In plaster walls, it can be laid in a shallow groove and plastered over.

• *Twisted indoor wire.* Twisted indoor wire may have two-, three-, or four-conductor insulated wire for remote-control wiring. It, too, comes in seven-strand, No. 20 AWG. The solid colors used for coding coincide with color codings of remote-control devices. The two-conductor wire, for instance, comes in blue and white, or in red and black. The three-conductor wire has red, white, and black conductors. Four-conductor wire has red, white, black, and yellow conductors. This wire may be installed in the same manner as flat wire. Fig. 10-12.

• *Flat, outdoor wire.* This wire is available in a three-conductor, neoprene-insulated jacket for low-voltage, remote-control applications. It comes in No. 18 AWG and may be used overhead or on underground applications. Fig. 10-13.

• *Multiconductor wire.* Number 20 AWG multiconductor wire is available in nineteen-conductor or twenty-six-conductor cable. It is used for making runs from master-selector switches and motor-master control units. The nineteen-conductor cable is

10-10. *A silicon rectifier. (GE)*

10-11. *Flat indoor wire, No. 20 AWG. (GE)*

10-12. *Twisted indoor wire, No. 20 AWG. (GE)*

10-13. *Flat outdoor wire, No. 18 AWG. (GE)*

231

made up of nine pairs of red and black wires, with one white common wire. The twenty-six-conductor wire has twelve pairs of red and black wires, plus one white common conductor and one blue wire for a locator lamp, or for a motor lead. Fig. 10-14.

• *Push-button master switch.* Where many circuits must be controlled from one convenient location, these master selector switches perform the necessary selection of only those circuits wanted. Or, by sweeping the *off* and *on* sides of these switches, a manual master control of all circuits may be used for one or more, or all eight of the switches. It may even be extended to other individual switches when desired. Fig. 10-15.

Also available is a "dial-type" master selector switch for twelve circuits, where each dial permits individual control; or by pressing and then sweeping all circuits, master control is accomplished. Both push-button and dial types are available with or without pilot lights. Both have a directory for locating individual circuits. Fig. 10-16.

• *Motor-master control units.* Motor-driven, master control units are used to turn *on* or *off* up to twenty-five individual circuits. When cascaded, the pressing of a single master switch can control any number of circuits, depending upon the number of motor-master units that are ganged together. Ganging is accomplished by having the last position of the first unit wired to start the second unit, the last position of the second unit wired to start the third, and so on. For automated lighting control, these units can be activated by program timers or photocell relays. Fig. 10-17.

Illustrated in Fig. 10-18 is the motor-master control unit starting switch, which is connected to lead (A) and white lead (B). When the starting switch is activated, the contact arm (C) starts to rotate in the direction indicated. The arm connects the white lead to contact points 1 through 25. It also picks up the slip ring connected to the white lead and keeps it completed. Complete revolution requires seventeen seconds.

Figure 10-19 shows a circuit that provides maximum flexibility of control. The circuit treats each floor as a unit for individual switch control. It uses master-selector control and motor master for turning off all floor relays. A central control also has individual control of each floor motor master through a master-selector switch. This switch has an eighth position control. This eighth position is a motor master that starts

10-14. *Multi-conductor cable, No. 20 AWG. (GE)*

10-15. *A push-button master switch. (GE)*

10-16. *A dial-type master switch. (GE)*

10-17. *A motor-master control. (GE)*

10-18. *Internal wiring of motor-master control. (GE)*

10-19. *This type of relay control center can be used in a two-story house and in the basement. The circuit provides maximum flexibility of control. It treats each floor as a unit for individual switch control, master-selector control, and motor-master control for turning off all relays. The eighth position controls a motor master that can start all the floor motor-master units. This circuit can be extended by adding a program clock and/or a photoelectric relay. (GE)*

10-20. *The component cabinet houses relays, motor masters, and power supply for ease of wiring. (GE)*

10-21. *A pilot light assembly for a master-selector switch. (GE)*

all the floor motor-master units. This circuit can be extended. Simply add a program clock and/or a photoelectric relay.

● *Components cabinet.* Cabinets for housing relays, motor masters, and power supplies make wiring easy. There are several sizes of cabinets available. Figure 10-20 will hold up to twenty-four relays, a twenty-four-volt transformer and rectifier assembly, two motor-master control units, and all other bus bars and terminal strips necessary for a neat-appearing, easily wired control center for the switching system. A steel separator is used to isolate the low-voltage section from the high-voltage section.

● *Pilot light assembly.* An assembly of twelve pilot lights mounted on a bracket fits behind the master selector switch. Each light illuminates the director tab corresponding to the circuit number indicated on the bracket. A plastic shield permits only a narrow slit of light to illuminate the printed tab. There are bracket mounts on the upper section of the two-gang plaster ring used for the master selector switch. Fig. 10-21.

● *Wall plates.* Wall plates are available to fit over the switches and give an attractive appearance to the wiring system. They are made of stainless steel or plastic. They range from a one-gang standard size with one vertical opening for standard remote-

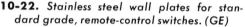

10-22. *Stainless steel wall plates for standard grade, remote-control switches. (GE)*

10-23. *Narrow-design switch plates for narrow partitions. (GE)*

control switches to a six-gang standard size with six vertical openings. Fig. 10-22. Figure 10-23 shows narrow-design wall plates for metal partitions where a minimum of space is available.

• *Extension switch.* This is a switch arrangement with a plug-in capability. It is designed to be used on top of a desk where you may want to control a number of remote operations without getting out of your chair. The cable, 8 feet in length, will give accessibility to anyone within that distance of a receptacle. Fig. 10-24.

• *Pilot lamp.* Figure 10-25 is a pilot-lamp bulb. It has a voltage rating of 24 to 36 volts. It is designed to last indefinitely unless it is damaged physically.

BASIC CIRCUITS. Some of the possibilities for remote-control applications of low-voltage relays and control switching are given in the next few illustrations. Figs. 10-26 through 10-33 (Pages 237-243).

10-24. *An extension switch and matching receptable. (GE)*

LOAD PILOT RELAY

LINE 5

LINE

5th FLOOR

LOAD PILOT RELAY

LINE 1

LINE

1st FLOOR

FLOOR SWITCHES

FLOOR SWITCHES

BASEMENT

RMC2PL

4 3 2 1

8 7 6 5

10-25. A replacement lamp for switches. (GE)

10-26. Multistory, separate transformers for master selector control of individual relays on different floors, where separate floor transformers and floor switches are also used. This circuit diagram shows the necessary wiring requirements. This circuit is especially useful for hallway lights controlled from one station. (GE)

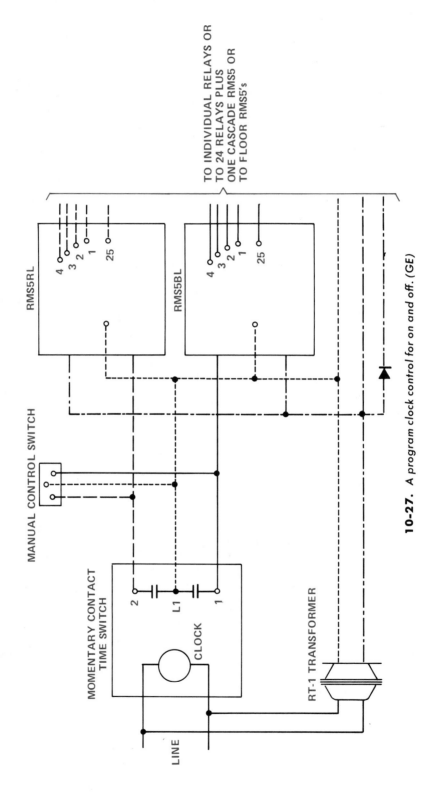

10-27. A program clock control for on and off. (GE)

10-28. A time switch, or photoelectric control, for on and off. (GE)

10-29. A thermostat-controlled relay for on and off. (GE)

CONTROL FOR ELECTRIC HEATING

10-30. *Selective switching using rectifiers. (GE)*

Your imagination is perhaps the only limitation to the application of low-voltage, remote-control switching.

Residential Installation Pointers

Install frames in relay boxes. Frames can be inserted only one way. Frames (brackets) for standard relays are installed in a similar manner. However, there are no bus bars for automatic connection to the relays, and wiring to each relay must be made through pigtails attached.

- *120-volt wiring.* This is installed in the usual manner, running the branch circuits to the 120-volt side of the relay boxes. Plug in the stripped conductors (strip ½ inch) into the pressure lock terminals. Make sure that the white and black wires are plugged into the terminals as marked. Only one set of terminals on each bus section is used for branch circuit supply. The other set may be used to extend, or couple, the two sections together. Next, make runs from relay boxes to the receptacles. Plug the ½-inch stripped conductors into the designated relay for that receptacle. Make sure that the white wire is pushed into the "white" pressure-lock terminal on the relay, and that the black conductor goes into the other terminal of the relay.

• *Mount switch supports.* Standard plaster rings for the thickness of the finished walls should be mounted at all switch locations. For best appearance, and for positioning the switches so that the words *on* and *off* appear in the normally accepted manner, the plaster rings should be mounted horizontally. Fig. 10-34.

• *24-volt wiring.* General Electric has a color-coded remote-control system that must be followed to make wiring installations easier. Follow the wiring diagrams. Remember that the motor-master control unit *must* be connected to the transformer *ahead* of the rectifier as it cannot run on rectified current (DC). It is an AC-operated unit. Use the color coding to eliminate possible errors in wiring.

The order in which the components are wired depends upon the preference of the contractor and in no way affects the speed of installation. However, the use of multi-conductor remote-control cables, with numbered as well as color-coded leads, cuts down on installation time.

Wires should be stapled, using tackers made for installing electric-heating wires and cables. Do not use flat, sharp-edged staples that cut into the insulation. Stapling guns, such as the Arrow Type T25 tacker, with ⅜-inch staples, or the Bostitch T5-8W tacker, are recommended. It is important to allow at least 12 inches of wire at the end of each run, placing a staple close to the plaster ring for the switch.

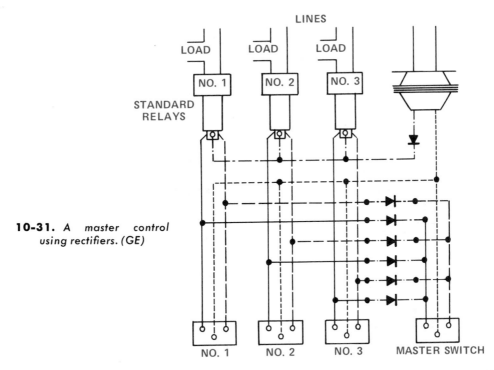

10-31. *A master control using rectifiers. (GE)*

10-32. A typical wiring diagram and layout for residential construction. (GE)

LINE-VOLTAGE SYMBOLS		LOW-VOLTAGE SYMBOLS	
Symbol	**Description**		
———	2-Cond. 120V Wire or Cable	— · · —	Remote-Control Low-Voltage Wire
—///—	3-Cond. 120V Wire or Cable	T	Low-Voltage Transformer
◯	Ceiling Receptacle	⟶⊢	Rectifier for Remote Control
◯̣	Floodlight	B$_R$	Box for Relays and Motor Master Controls
⊏—◯—	Valance Light	R	Remote-Control Relay
—©—	Clock Receptacle	R$_P$	Remote-Control Pilot-Light Relay
Ⓛ	Keyless Lampholder	P$_{11}$	Separate Pilot Light, R.C. Plate
Ⓛ$_{PS}$	Pull Chain Lampholder	P$_{10}$	Separate Pilot Light, Inter. Plate
⊏⊖	Double Receptacle, Split Wired	MS	Master-Selector Switch
⊖	Grounding Receptacle	MM$_R$	Motor Master Control for ON
⊖$_{WP}$	Weatherproof Grounding Receptacle	MM$_B$	Motor Master Control for OFF
⊖$_R$	Range Receptacle	S$_M$	Switch for Motor Master
—▲$_{CD}$	Clothes Dryer Receptacle	S$_{F6}$	R.C. Flush Switch
S$_L$	Lighted-Handle Mercury Switch	S$_{F7}$	R.C. Locator-Light Switch
S$_P$	Push-Button, Pilot Switch	S$_{F8}$	R.C. Pilot-Light Switch
S$_D$	Closet Door Switch	S$_{K6}$	R.C. Key Switch
		S$_{K7}$	R.C. Locator Light Key Switch
		S$_{K8}$	R.C. Pilot-Light Key Switch
		S$_{T6}$	R.C. Trigger Switch
		S$_{T7}$	R.C. Locator Light Trigger Switch
		S$_{T8}$	R.C. Pilot-Light Trigger Switch
		S$_{T4}$	Interchangeable Trigger Switch, Brown
		S$_{T5}$	Interchangeable Trigger Switch, Ivory
		⊖$_{RO}$	Remote-Control Receptacle for Extension Switch

10-33. *Suggested symbols. (GE)*

Remote Controll

10-34. *A plaster ring, mounted horizontally, is used to support one, two, or three flush switches. (GE)*

Adhesive tape, wrapped around the wires, can be numbered to correspond with the relay to which they are connected. Multi-conductor cables are already numbered and need no further identification, except perhaps where more than one master selector switch is to be installed in the same location.

IMPORTANT SAFETY PRECAUTION. No electrician would ever think of bunching up bare line-voltage conductors in an outlet box during the plastering operations without first removing the fuses. Similarly, the 24-volt connections from a remote-control transformer *must be disconnected* before bunching up the low-voltage wires at the switch locations. This must be done to prevent workers from shorting the wires in an attempt to get power or light. Standard grade relays are not designed to operate on extended energizing periods that may cause relay burnouts.

CHECKING THE INSTALLATION. Every new wiring installation, including remote-control wiring, should be checked as soon as the power is turned on.

To check the system, each circuit should be tried from each switch point, including the motor-master control switches, to insure that everything is operating correctly. If care was taken in following the installation procedures, as outlined, the system should work perfectly.

The following suggestions are offered as a guide to the proper procedure for locating and correcting installation errors.

- *Entire system dead when power is applied.*
1. Check for power at transformer primary.
2. Check voltage of secondary.
3. Check to see is transformer is dead. If there is power at the primary leads (120-volt leads), short out *secondary* terminals with screwdriver. A spark indicates transformer is "alive." NOTE: This procedure does not harm the transformer if done only momentarily to produce a spark. Look for a short or open circuit in the transformer loop wiring as follows:

a. If there are separate zones for the location of relay-center boxes, or if there are several transformer wires to groups of relays mounted in outlet boxes, disconnect all but one.

b. Check to see if switches will work in the zone or relay center left connected. Connect each zone branch to the transformer and check to find the one branch that is shorted.

4. When the zone is found in which the relays fail to operate, look for a short or open circuit in the loop circuit. This may be done by disconnecting the transformer connections at the relay and checking with a "ring out" set (or continuity tester of any type) from location to location until the short or open circuit is found.

- *Individual circuit stays on or off — relay hums and the relay barrel becomes warm.* This fault is caused by a continuously energized relay coil.
1. Check to see whether an individual switch, or the switch on the nine-position master selector, is binding with the wall plate, preventing it from returning to the neutral position. (If the switch is not free to move, it may be held in either the *on* or *off* position and thus heat up the relay.)
2. Check for a shorted control switch by disconnecting the wires and shorting first the red wire to the white wire, then the black wire to the white wire. If this operates the relay, then the control switch should be checked and replaced if necessary. It may be

necessary to allow the relay to cool before making this test if the coil has been energized for an extended period of time.

3. Inspect the relay for possible short circuits between conductors at low-voltage soldered connections at the end of the cap, near the point where the wires are connected.

4. If trouble persists, check relay.

5. If this check does not correct the trouble, there is a short circuit in the control wiring between relay and control switches. This may be found in much the same manner as described before for the transformer loops.

• *Individual circuit stays on or off—relay does not hum or become warm.* This fault is caused by an open circuit in the control wiring.

1. Inspect the connections at the low-voltage end of the relay to be certain there are no open circuits.

2. Inspect connections at the control switches and master-selector switches to see that they are not broken and are making good contact.

3. If fault persists, check for continuity in the control wiring.

• *Circuit turns on when control switch is turned off.*

1. Reverse connection to the red (on) and the black (off) terminals of the relay or switch. This error will not occur if color-coded wire is used and colors are matched up in making the connections.

LOW-VOLTAGE WIRING

Built-in Home Entertainment, Intercoms, Door Chimes and Openers, Climate Control, and Burglar Alarms

An electrician is called upon to install a variety of electrical devices. Most home entertainment systems require a low-voltage wiring installation in addition to a 120-volt circuit. Some multipurpose systems can operate doorbells, intercoms, a source of music, or a burglar alarm, as well as smoke and fire detectors. Much of the wiring for such a system should be done before plaster or wallboard is applied.

• *Entertainment and intercoms.* Figure 10-35 is an example of a communications center for the home. It requires six-conductor wire, 14/2 Romex, and two-conductor No. 18 or 16 wire. A "communi-center" radio intercom with digital clock and tape player is available. A record player may be added. This unit can have a complete series of door, patio, and inside speakers.

• *Door chimes and openers.* Doorbells are the most common of the low-voltage circuits encountered by an electrician. Fig. 10-36. In this circuit the transformer supplies 10 or 16 volts, depending on the design of the bell (or buzzer, or chime). Fig. 10-37. The push button energizes the buzzer or bell. Figs. 10-38 and 10-39. The transformer may be mounted on or near the house distribution-panel box.

In some instances a low-voltage door opener is necessary to allow access to a room only upon a button being pressed from a remote location. This is usually an installation for a store, bank, or industrial plant where controlled access to a specific area is needed. Fig. 10-40.

• *Climate control.* Most thermostats require low-voltage controls, usually 24 volts. The wire from the thermostat to the furnace is two-wire No. 18. Fig. 10-41. This wire

10-35. *A block diagram of a communications center for the home. (NuTone)*

10-36. *A door bell circuit.*

10-37. *A door bell or chime transformer.*

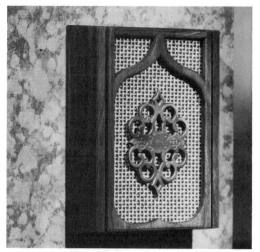

10-38. A chime. One sound or note for rear entrance and two notes for the front entrance. (NuTone)

10-39. Push buttons.

10-40. An electric door release. This flush-mounted mortise type installs in place of a lock strike plate on wood, or on a light-gage metal door jamb. The entrance remains secure unless it is opened with a lock key, activated by a lock-release control on an apartment speaker, or opened by a special postal lock provided within the main-entrance panel. It connects to a 16 VAC, 1-ampere source. (NuTone)

TWISTED THERMOSTAT CABLE

No Jacket — Plastic Insulated Conductors

PLASTIC JACKETED THERMOSTAT CABLE

Use: For thermostat controls in connection with heating and air conditioning installations.

10-41. Thermostat wire. Used for thermostat controls in heating and air-conditioning installations. Also used for bell and annunciator systems. Designed for use on low-voltage systems. Solid copper conductors, thermoplastic insulation, color coded, conductor twisted, overall jacket of plastic.

may also be used for bell wiring or other low-voltage applications. Do not run low-voltage and 120-volt wiring in the same conduit, however. Thermostat wire is also available in three-conductor, four-conductor, and five-conductor cables. It usually has solid copper conductors, is thermoplastic insulated and color coded, and is twisted with an overall plastic jacket. Use insulated staples, being careful not to break the insulation on the cable when pounding the staple in place.

Air-conditioning units require more conductors than a simple furnace control. If the furnace and air-conditioning units are combined for a central air-conditioning and heating capability, it is probable that the cable should be five-conductor. This specification varies with the manufacturer; so be sure to check the unit for specific directions before installing the wire. Fig. 10-42. Remember also that this low-voltage wiring is *only* for the control circuits. The 120- or 240-volt wiring is separate.

● *Burglar alarms, smoke detectors, and fire alarms.* Because burglar alarms vary in design, you should know exactly which unit is being installed before beginning to wire the switches and detectors. Figure 10-43 shows a typical wiring diagram for a device to detect forced entry, as well as smoke and fire. Note the different voltages and wire sizes needed to get the job done properly within the local and national codes.

LIGHTING CONTROLS

Controlled light is appropriate, wherever people get together, to provide suitable lighting for the occasion.

Controls For Incandescent Lamps

TRANSFORMERS. Light controls using transformers are available for a number of applications and wattages. Transformers (autotransformers to be exact) are designed to completely control light intensity by controlling the voltage applied to lamps. Rotating the knob moves a brush contact over a commutator. This produces any desired intensity of light, ranging from complete darkness to full brightness. Because the brush is always in contact with the commutator, operation is smooth, silent, and flickerless. Only the amount of power required to produce the desired illumination is used. Operation is always cool, safe, and economical, unlike resistance-type controls that regulate by dissipating power in the form of heat.

Transformers are made for a number of loads and voltages. Fig. 10-44. They can be installed in a wall box and protected according to the NEC requirements. Figure 10-45 shows a transformer for either home or commercial use. Note the thermal overload protector. Wall plates are used to give a finished appearance.

10-42. *Although this is a four-conductor cable, there are available to the electrician almost any number of conductors. They come color coded or in pairs, with or without shielding.*

Security/Fire System Wiring

10-43. *A fire and security alarm system. (NuTone)*

When replacing an existing wall switch with a light control transformer, it may be found that a "switch loop" has been used between the lighting load and the switch, as shown in Fig. 10-46A. This usually does not present a problem because the dimmer can be incorporated by changing the circuit according to Fig. 10-46B.

If on/off control from a number of locations is desired, a three-way wall switch may be installed in the input line to the dimmer to turn the lights on or off from more than one location. Fig. 10-47. When turned on, the lights will be at the brightness level previously set on the dimmer knob. Other variations are possible that will meet the requirements of the installation.

ELECTRONIC DIMMERS. Electronic dimmers are used to control incandescent lamps. Figs. 10-48 through 10-51. These dimmers are small enough to fit into a standard wall box. The wall plate can be any standard type used for a toggle switch. Fig.

COIL

10-44. *An autotransform-*
er. (Superior)

INPUT
SWITCH

BRUSH
ARM

THERMAL
OVERLOAD
PROTECTOR

WIRING
LEADS

10-45. *A typical diagram for a*
transformer-controlled light. (Su-
perior)

BLACK

SWITCH
OPERATED
WITH KNOB

L

INPUT
120 VOLTS
AC

WHITE

RED

THERMAL
OVERLOAD
PROTECTOR

OUTPUT TO
LAMP LOAD
0-120 VOLTS

N

10-46. *(A) When replacing an existing wall*
switch with a light control, it may be found that
a "switch loop" has been used between the
lighting load and the switch. (B) This usually
does not present a problem because the dim-
mer can be incorporated by changing the cir-
cuit. (Superior)

SWITCH

SWITCH LOOP
CABLE

LAMP

Light Control

OLD
CABLE

NEW
CABLE

LAMP

LINE
120 VOLT LINE
NEUTRAL

JUNCTION
BOX

A

LINE
120 VOLT LINE
NEUTRAL

JUNCTION
BOX

B

10-47. Control from multiple locations. (Superior)

10-48. Electronic control fits into the switch box. (Superior)

10-49. Wall plate and knob mounted on the electronic control make a neat appearance. (Superior)

10-50. An electronic light control, mounted in a box. (Superior)

10-51. A three-way electronic light control with wall plate and knob. (Superior)

10-52. Note that the wall plate and knob fit in place the same way that a regular wall switch goes into a box. (Superior)

10-53. A hi-lo switch. (Slater)

10-54. Snap-off fins on electronic dimmers make it easy to fit into almost any combination of wall plates. (Slater)

10-52. These units have an SCR (silicon-controlled rectifier) that controls the voltage by the use of a small variable resistor. This potentiometer in turn controls the resistance the SCR puts into the circuit. These are also available in a three-way configuration.

The switch in Fig. 10-53 has a "hi-lo" type of control with "center-off." ("Hi" is up, "center" is off, and "lo" is the down position of toggle switch handle.) A diode is usually placed into the circuit for the "lo" position, with full voltage on the lamps for the "hi" position. They may be mounted in the vertical position with the "hi" at the top or in a horizontal position. However, the plate reads correctly only in the vertically mounted direction.

Some rotary dimmers have breakaway fins that allow for fitting into the space available. The fins dissipate the heat generated by the electronic circuitry. Fig. 10-54.

PHOTOELECTRIC-CONTROL SWITCHES. Photoelectric switches are available for use in automatic circuits, for the person who does not want to be bothered with remembering to turn a certain light on at dusk and off at sunrise. Fig. 10-55.

"Nite-on" photoelectric switches operate by turning lights on at dusk and off at dawn, automatically. They have a built-in delay in each unit to prevent flickering or

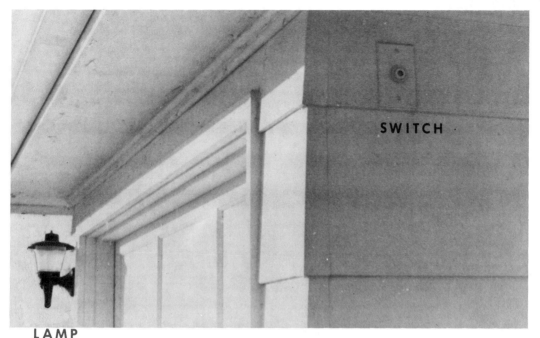

SWITCH

LAMP

10-55. *A photoelectric control, mounted facing north, controls the lamp automatically.*

flashing of lights. Flickering might be caused by sudden extraneous light sources such as headlights or flashlights.

A wiring diagram for a photoelectric switch is shown in Fig. 10-56. The photocell is a sealed unit made of cadmium sulfide. It will operate at −40 to 71°C [−40 to 160° F.]. Most will operate 1000 watts of tungsten lamps or 450 watts of mercury-vapor lights at 120 volts AC. Photocells have a rubber gasket that weatherproofs them when they are mounted in a standard metal or insulated box.

10-56. *Wiring diagram for an automatic, outdoor weatherproof switch, controlled by light. (Slater)*

WIRING DIAGRAM

NITE-ON

BLACK   BLACK   RED   BLACK

120V AC   WHITE   LIGHT OR LOAD

WHITE   WHITE

10-59. *A momentary-contact switch for motor-driven, remote-controlled light controllers. (Superior)*

10-57. *Power consumption of dimmer floures-cents. (Superior)*

SINGLE-LAMP DIMMING BALLAST

TWO-LAMP DIMMING BALLAST

10-58. *Single-lamp and two-lamp dimming ballasts with associated circuits. (Superior)*

10-60. *A positioner controller. (Superior)*

Fluorescent Lamps

Controls have been designed to permit fluorescent lamps to be dimmed. These electronic dimmers will control fluorescent lamps over a wide range of brightness. The lamps will start quickly at any intensity setting, and will operate without flicker even at low or high settings. Models rated for use with up to 120 lamps are available. Lamps must be 40-watt, rapid-start fluorescent and equipped with *special dimming ballasts.* White or colored lamps may be used. No dimming auxiliaries are needed. The dimmers incorporate filters for suppression of radio-frequency interference (RFI). They need no additional chokes, as was formerly the case. (A choke is a coil of wire or inductor.) This type of dimmer is somewhat expensive for residential applications.

POWER CONSUMPTION. Operation of fluorescent lamps at reduced intensity provides substantial cost savings because power consumption is also reduced. Fig. 10-57. The curve of light output, contrasted with input current and watts, shows the decrease in power supplied to the ballasts as the lighting is varied from full intensity to blackout.

Light-control models are offered in self-contained and remotely controlled varieties. *Self-contained* models mount on the wall. They are offered in maximum capacities of 10 to 30 lamps. *Remotely controlled* models can be mounted against any flat surface. They require a separate controller which can be located up to 200 feet from the dimmer. Maximum capacities of 30 to 120 lamps are available. These are designed for commercial and industrial uses.

LAMPS. Multiple lamps for fluorescent fixtures should be the same size and age. Ideally, the lamps should have approximately 100 hours burn-in time to reach maximum efficiency. Lamps should not be installed in a cold draft, either natural or from air-conditioning outlets. Cold lamps become unstable and will not dim satisfactorily. Fig. 10-58.

Motor-controlled remote dimmers are available. A momentary switch is used for control, or a positioner controller may be used. Figs. 10-59 and 10-60.

Self-contained models can be combined with three-way switches to allow lights to be turned on and off from either of two locations. An added advantage of this method is that it is not necessary to disturb the intensity setting when the lights are turned off. The lights will then be at the same intensity when turned on again.

A bypass switch allows the lighting load to be quickly transferred from the dimmer output to the line. A two-pole, double-throw switch is required. The switch can be placed at the control station or at a remote point. Fig. 10-61.

10-61. *A bypass switch with remotely controlled series dimmers. (Superior)*

QUESTIONS

1. What is a relay?
2. What is the advantage of low-voltage wiring?
3. What is the voltage used on low-voltage control systems?
4. How is low voltage obtained?
5. What is a rectifier?
6. Where are rectifiers used in low-voltage systems?
7. What size wire do you use in a low-voltage system?
8. What is meant by a multiconductor cable?
9. What is a motor-master unit?

10. How do you install the wires in a low-voltage system?
11. What would you check if the low-voltage system did not operate correctly when you finished hooking it up? What would you check first?
12. What type of wiring is used with thermostats?
13. What are two ways of controlling light?
14. What is a SCR?
15. Can fluorescent lamps be dimmed?
16. What is a by-pass switch?

Wiring with BX, EMT, Flexible, Plastic, and Rigid Conduit

There are a number of methods used to enclose wires for houses. The primary objective is to make sure the electrical wiring and its many points of junction or connection cannot cause a fire under normal operating conditions.

As a rule, conduit is not used for wiring houses. It is of great use in industrial and commercial locations, but becomes rather expensive for use in homes. The metal of the conduit protects the wire insulation and the wires themselves from physical damage that may result in shorts or opens. These shorts may cause overheating and fire. Arcing can cause explosions in some atmospheres. The enclosed conduit type of wiring can be installed to prevent explosions caused by arcing. In some locations, it is necessary to enclose totally a switch or outlet to prevent an explosion. Hospitals, for instance, need special attention paid to the operating rooms. Here, gases used for anesthetizing patients may cause an explosion if ignited by a spark.

Some of the applications of conduit to industrial and commercial locations will be shown here for your information. You may want to add a special wiring arrangement if you have a special-purpose room in a home you are building or remodeling.

BX CABLE

BX cable is the name applied to armored or metal-covered wiring. It is sometimes used in home applications. BX (the letters stand for metal-covered, flexible wire) sometimes meets the need in home applications for flexible wiring. It is used to connect an appliance, such as a garbage disposal unit, which vibrates or moves a great deal. For example, a garbage disposal unit is installed under a sink in the kitchen. The unit will vibrate when in normal use. This vibration calls for a flexible conduit or wire. The BX is attached to a junction box on the wall and through a special connector to the disposal unit.

BX, however, presents some rather unique problems. It must be cut properly. If it is not properly cut, the insulation covering the wires inside may be shorted to the protective metal covering. There is a fiber bushing designed to fit into the cut end of the BX. This bushing insulates the sharp end of the BX (where it was cut) and prevents shorts.

BX Cutter

A tool has recently been designed to cut BX without the ragged edges produced by a hacksaw. Fig. 11-1. It is very efficient and can be carried on an electrician's belt with other tools.

USING THE CUTTER. Once the BX cable has been secured in the cutter, with the thumbscrew on the bottom, squeeze the handle. Squeezing the handle overcomes the

11-1. *BX cable ripper. (Seatek)*

force of the built-in spring. At the same time, it lowers the circular cutter against the BX casing. Do not apply too much pressure if you want a fast cut. At the point when the force required to turn the handle suddenly reduces, even when you increase the squeezing force, you have reached the pressure limit. Stop and release the tool. Release the hand pressure and turn the crank slightly. The tool should snap back to its original position for easy removal from the BX.

After the tool has been removed from the BX, hold the cable on each side of the slit and twist the casing counterclockwise until the casing separates. The insulated wires inside the BX casing are now exposed. If this separation doesn't happen easily, you haven't cut the BX completely. You will therefore have to go back and insert the tool again to finish cutting the armored cable housing.

Once the BX has been cut, you must install it properly in a box for termination. There are a number of connectors for this type of job. Fig. 11-2.

FITTINGS

Connectors

• *Box connectors* are designed to secure cable, metallic or nonmetallic, to junction enclosures. These connectors are not watertight and are used indoors. Fig. 11-2.

They are available in straight, 45°, and 90° angle styles. Some types have a screw-clamp method of gripping the cable; others utilize a squeezing principle. Figs. 11-3 and 11-4. Their purpose is to provide a clamping action without causing injury to the cable. The clamping action prevents connected wires from getting pulled apart by stress put on the cable from outside the enclosure. Fig. 11-5.

Box connectors used with metal-clad cable provide a continuous ground through the cable armor to the metal box. The cable is attached to the box with a connector. This assures a good ground connection. Box connectors are easy to install. Insert them in a knockout or drilled hole and secure them in place with a locknut. Many out-

TWO SCREW
CLAMP TYPE

SINGLE SCREW
CLAMP TYPE

SQUEEZE TYPE

A

ROMEX

SERVICE
ENTRANCE

CORD OR
BARE GROUND

B

11-2. *Box connector clamps. (Appleton)*

C

90°- 45° FOR
BX CABLE

DUPLEX FOR
BX CABLE

STRAIGHT FOR
BX CABLE

11-3. *BX cable connectors. (Appleton)*

11-4. *A squeeze-type connector used on Romex. This type cannot be used on BX.*

LOCKNUT

ROMEX CABLE

WALL

11-5. *A straight, two-screw, clamp-type connector used for Romex. In some cases this type of connector can be used with BX. Note difference in Fig. 11-7a.*

let and switch boxes have built-in cable clamps. Fig. 11-6. When such clamps are provided, no box connector is needed.

Included in the box connector series is a straight duplex connector. This will take two metal-clad cables. All other connectors will take only one cable. The 45° and 90° types have removable hoods to allow easy insertion of the cable. Fig. 11-7.

Box connectors are used in all types of construction where the use of metallic and nonmetallic cables and raceways is permitted. They are also applied to machines that have a light spray of oil or other liquids. Flexible cables are practical in most tight-area wiring. Fig. 11-8.

• *Ninety-degree knockout box connectors* are designed for use in connecting conduit to outlet boxes. They may be used on other similar devices. Two types are available—threaded and no-thread. Fig. 11-9.

• *Cable connectors (watertight)* are used for entrance of flexible cords and cables to electrical equipment. Fig. 11-10. They provide clamping action and reduce cable-

11-6. *An outlet box with BX clamps built-in. (Appleton)*

SQUEEZE   SCREW

WALL

LOCKNUT

11-7a. *Forty-five degree and ninety degree connectors for BX cable. (Appleton)*

REMOVABLE COVER

BX CABLE

WALL

11-7b. *A box connector with BX clamp. Note the removable cover for ease in placing the wires into the box. (Appleton)*

SCREW CLAMP

WALL

CORD OR BARE GROUND WIRE

11-8. *A connector holding a cord in a box. (Appleton)*

11-9. *Ninety-degree knockout elbows. This one has a conduit fitting and attaches to the box with a locknut. (Appleton)*

11-10. *Watertight cable connectors. These are special application connectors for use in special areas in residential buildings. They are used mostly in commercial and industrial applications. (Appleton)*

insultion abrasion and wear. Watertights seal out vapor and dirt. They are both watertight and oiltight. Thus, they help prevent deterioration of valuable enclosed mechanisms. Their use can also prevent wire terminal strains. Strains result in dangerous broken connections. Use of watertights can prevent short circuits due to abrasion at the cable connector.

These connectors are made of light-weight aluminum. They consist of four parts: *body, grommet, ring,* and *cap.* An oil-resistant *neoprene* grommet provides a tight seal and firm grip on the flexible cord, or cable. The ring under the clamping cap allows the grommet to be compressed tightly around the flexible cord, or cable, without friction on the cap.

Watertight connectors are available in straight and 90° angles. Hub sizes are $\frac{3}{8}, \frac{1}{2}, \frac{3}{4}$, 1, and $1\frac{1}{4}$ inch. A hub is an enlargement on the end of a piece of pipe or conduit. This means another pipe or conduit of the same size may be inserted and sealed with a satisfactory joint. In this case, a hub means that the end of the connector is such as to accept a piece of conduit of the size specified on the connector. Installation of watertight connectors is easy. Fig. 11-11. These connectors have standard, tapered pipe threads for installation directly into a threaded cast fitting. A locknut may be added if the cable is to be used on an enclosure that is not threaded. This may occur when used on an outlet box or a steel junction box.

11-11. *Watertight connectors in a unilet. The unilet is usually associated with conduit installations. In some home installations this can be a very useful arrangement. (Appleton)*

FEED CABLE ALL THE WAY
INTO THE WIRING CHAMBER

CABLE

BEFORE ASSEMBLY   AFTER ASSEMBLY

BX CABLE AND THE CODE

BX cable has an official designation of AC or armored cable. This type of cable is now covered by its own article in the Code. It has been drawn out and separated from the nonmetallic cables or Romex. This type of cable, according to Underwriters' Laboratories, has a copper or aluminum bonding strip in contact with the armor. The covering is made of steel. There are insulated conductors inside the cable for current carrying purposes.

Armored cable can be obtained in two-, three-, or four-conductor sizes varying from No. 14 to No. 1. All these will conform to the UL specifications. It is for use at temperatures of 60°C [140° F.] or 75°C [167° F.], depending upon its insulation over the conductor. This type is used for voltages up to 600. Other metal-covered cables are not covered by this article of the Code. You should read the cable description on the box when purchasing.

One of the requirements of the Code for BX is the insertion of an anti-shorting bushing, sometimes called a "red head."

Types of AC Cables

Certain letters are used to designate the type of cable with UL data on the conductors:

The letters ACT are used to indicate armored cable that has conductors with thermoplastic insulation. This type of insulation is usually designated with a T. Therefore, ACT designates armored cable with thermoplastic insulation.

AC indicates armored cable with conductors that are covered with rubber insulation of the grade specified by the Code.

ACH indicates armored cable that has rubber insulation that will withstand heat within the 75°C [167° F.] range.

ACHH indicates cable with a temperature range of 90°C [194° F.].

ACU indicates armored cable with rubber insulation, but of the latex grade.

Whenever L appears as a suffix, it indicates that lead has been used as a covering over the conductors.

Uses of Armored Cable

Armored cable is used for branch circuits, feeders, and light circuits. You have to be careful with the wiring of BX because it is possible for induction heating to cause problems. Induction heating takes place when the current feeding a particular device comes though a conductor in one metal-covered wire and returns to the source in another metal-covered wire inside a parallel arrangement of the two metal covers that also serve as conductors. Fig. 11-12.

INDUCTION. Induction takes place whenever a magnetic field cuts across a conductor. In this case, the magnetic field is established by the current-carrying wires. If the current flows as shown in Fig. 11-12A, then the magnetic field causes a path for an induced current in the metal BX covering and the junction boxes. This can be sufficient, at 30 amperes and over, to reduce severely the voltage available at the lamp. It may also generate heat that can cause a fire in anything contacting the heated metal covering or armored case.

Figure 11-12B shows the normal condition of wiring in conduit or BX. Both conductors are housed in the same metal jacket. The current flowing to the lamp and the current flowing away from the lamp are in different directions. They have a tendency to cancel the magnetic field generated by the current-carrying wires.

Type AC is used for signal and control circuits. Some installations of speakers in remote locations are common. It is used in new construction and in remodeling. However, it is more expensive initially than Romex. This factor alone limits its use in home construction and wiring.

Type AC or even ACL cable cannot be used where it is buried directly in the earth.

11-12. *Induction can be caused if the wiring is not properly pulled through a conduit or BX.*

Armored cable must be secured by approved staples, straps, or fittings. The straps or staples should not be spaced more than 24 to 30 inches apart. If the wire is concealed, you can support it every 4½ feet with staples or an appropriate fastening device. It should be secured within 1 foot of each outlet box or fitting.

Armored cable should be supported every 2 feet. This applies in particular to lamp fixtures or any enclosure or equipment. However, Greenfield flexible tubing and liquid-tight tubing can be left for as much as 6 feet without support. This does not apply to BX. BX must be supported every 2 feet.

Whenever you use another wiring method and it connects to a BX cable, the junction box must be a standard box. Wires must be spliced in the junction box.

11-13. *When changing from BX to conduit or some other type of wiring method, you must use a junction box for the transfer.*

INSTALLING ELECTRIC SERVICE WITH CONDUIT

Conduit is not too hard to install. It requires a conduit bender, hacksaw, screwdriver, hammer, and pliers. In some areas, the Code requires conduit for home use. This varies from place to place. Thus, it is best to check with your local power company to make sure. It is the most expensive type of wiring. If the building has more than three floors above grade level, Romex or nonmetallic cable cannot be used. Other types must be used—either BX or conduit.

Wires from the utility company's pole to your building are called a service drop. This is usually furnished by the company. These wires must be high enough to provide proper clearance above grade. They must not come within 3 feet of doors, windows,

Attachment must be located so service drop clearances can be achieved.

18″ min

Minimum is 18 ft over public streets, alleys, roads, and driveways on other than residential property.

Minimum is 15 ft over commercial areas, parking lots, agricultural or other areas subject to truck traffic.

Minimum is 12 ft over residential driveways and community areas such as parking lots and drive-in establishments not subject to truck traffic.

Minimum is 10 ft above finished grade sidewalks or from any platform or projection from which they might be reached.

Grade

11-14. *Various clearances for a service drop. (NMP)*

fire escapes, or any opening. The structure to which the service drop is fastened must be sturdy and able to withstand the pull of ice, wind, and other weather conditions.

Figure 11-14 shows some of the clearances for a service drop. These are in accordance with the National Electrical Code. Figure 11-15 shows how the required clearance is figured. In some cases, there are exceptions. Figure 11-16 shows one of the exceptions. This one is brought about by the length of the overhang.

Figure 11-17 indicates some of the details that must be considered when installing a service drop.

11-15. *Mounting the service head on the house.*

Service-Drop Conductors, 300 Volts Or Less

18" min

4' max

Service Riser Extending Through Roof

Meter

Service Drop Not Over 300 Volts

Required Clearance

18" min

48" max

11-16. One exception to the Code requirements for the location of a service drop.

11-17. *Proper installation of a service drop and mounting the weatherhead.*

Support For Triplex Aerial Drop

Weatherhead/Adapter Clamp Or Threaded

2 in. Or Larger Rigid Metal Conduit

Seal

Flashing

Note: Consult local authorities for installation data for mast services.

Mast Clamp

Roof Plate

Guy Fittings Where Backguying Is Required

Mounting Clamps With 1/2 in. Bolts

Offset Reducer For Connection

1/2 in. Bolts Run Through Studs

Bolt And Nut Flush With Inside Stud Face (Countersink)

Installation Procedures

Attach the service head to the building as suggested by the Code and your local regulations. Connect the conduit as shown in Fig. 11-18. Use a metal strap every 4 feet to fasten the conduit to the building. Connect the meter with conduit connectors. Use an entrance ell to turn the conduit into the house. The ell has two threaded openings corresponding to conduit size. Use an adapter to fasten the conduit into the threaded opening at the top of the ell. Into the lower opening fasten a piece of conduit

White

Black

Red

These three lead
wires must extend
out of entrance
head at least 36 in.

Rigid Conduit

Conduit
strap

11-18. *Using conduit for a service entrance. (Sears)*

Conduit
connector

Meter
socket

Rigid or
thin-wall
conduit

Conduit
adapter

Entrance
ell

Conduit
connector

Service panel

11-19. *Entrance cap, meter, entrance ell, and service switch. (Appleton)*

11-20. *A 90° female elbow. (Appleton)*

to run through the side of the house. Use a connector to attach the conduit to the electric service panel. Fig. 11-18.

FITTINGS. The following fittings are used.

• *Entrance elbows* are used customarily where the conduit enters a building. The size of the conduit and fittings depends on the size of the entrance wires.

Entrance elbows must be raintight. Before pulling the wires through the conduit, remove the cover on the ell. (*Ell* is short for elbow.) Use a fish wire to pull the wires into place. After connection of the wires, replace the cover. Entrance ells come in almost any shape and size to fit almost any installation. Fig. 11-19.

• *Ninety-degree female elbows* are designed for use with conduit-to-conduit, right-angle installations. Elbows are generally used instead of bending the conduit itself. Fig. 11-20.

• *Ninety-degree bushed elbows* are used to make right-angle installations, either conduit-to-conduit, or conduit-to-box. Fig. 11-21.

• *Pulling elbows* provide a simple yet effective means of pulling wires through raceways. They are used in cases where it is necessary to run the conduit around 90° corners. Figs. 11-22 and 11-23. The pulling elbows illustrated have removable covers which, when removed, allow you to pull the wires freely all the way through one side of the fitting, then through the other side. The cover is then replaced.

• *Forty-five female conduit elbows* are designed for use when a conduit-to-conduit, 45° angle installation is required. These elbows are generally used instead of bending the conduit. Fig. 11-24.

• *Steel-cupped reducing washers* are used with outlet boxes. They reduce the diameter of a knockout to a smaller opening. They are made of cadmium-plated steel and are used in pairs. One is placed inside the outlet box, the other outside. Conduit is placed through them with a locknut on either side. Fig. 11-25. Tighten the locknuts to

90°

ELBOW

THREADED CONDUIT

11-21. *A 90° bushed elbow. (Appleton)*

THREADED CONDUIT

THREADED CONDUIT

11-22. *A 90° pulling elbow. (Appleton)*

COVER AND GASKET REMOVED

WIRES PULLED AND RUN AROUND CORNER

THREADED CONDUIT

11-23. *A pulling elbow. (Appleton)*

45° FEMALE CONDUIT ELBOW

THREADED CONDUIT

THREADED CONDUIT

11-24. *A 45° female conduit elbow. (Appleton)*

REDUCING WASHER

3/4" KNOCK-OUT

1/2" THREADED CONDUIT

LOCKNUT

1/2" LOCKNUT

BOX WALL

REDUCING WASHER

11-25. *Steel cupped reducing washers. (Appleton)*

complete the installation. A ledge on the edge of the washer prevents it from slipping after it has been installed.

• *Entrance caps for use with threaded rigid conduit* are also known as service heads, entrance fittings, entrance heads, or weather heads. They are used to bring electrical power into a building. Current-carrying wires are brought from the outdoor lines to the entrance cap. The entrance cap is anchored on the building. Wires are then pulled from the service switch inside the building and through the conduit. The conduit passes from the inside wall to the outside wall of the building. The wires are pulled into an entrance elbow mounted on the exterior wall of the building. Fig. 11-26.

Conduit comes out of the building into an ell. Then conduit is connected from the ell (located at the bottom of the building or foundation) to the service entrance cap. Wires are pulled from the ell, through the conduit into the entrance cap. They are then threaded through the wire holes in the insulating block of the service cap. The wires should be extended about 18 inches [455 mm] and cut off. At this point, the interior circuit wires are spliced to the exterior wires.

Several types of service caps are available. All have removable covers, in which a detachable insulating block keeps the individual circuit wires separated. Some have snap-on covers that eliminate attaching screws on the cover.

In some cases, service-entrance cable is used instead of conduit. For these installations, the cable entrance caps have a clamp neck instead of threads or setscrews. The

TO OPEN — PLACE SCREW-DRIVER BETWEEN INSULATOR AND CENTER OF COVER . . . TWIST . . . AND IT OPENS.

TO USE — SLIDE FITTING HUB ONTO E.M.T. OR RIGID CONDUIT, FASTEN TIGHTLY WITH ALUMINUM SETSCREW AND THREAD WIRES THROUGH HUB AND INSULATOR.

TO REASSEMBLE — RESEAT INSULATOR, SLIDE COVER ALONG GUIDES AND SNAP INTO POSITION. THE INSTALLATION IS COMPLETE!

11-26. *Entrance caps for use with threaded rigid conduit. (Appleton)*

SNAP-ON ENTRANCE CAP

ALUMINUM THREADED REA WITH FLANGE

MOUNTING HOLE

CLAMP

TWO HOLE PIPE STRAP

SERVICE ENTRANCE CABLE

METER BOX

BUILDING WALL

BOX

ENTRANCE ELL

11-27. *An entrance cap. (Appleton)*

11-28. *An entrance cap with compression ring for EMT. (Appleton)*

INSULATOR

REMOVABLE COVER

BODY

COMPRESSION RING

COMPRESSION NUT

SNAP-ON CAP

INSULATOR

WIRES

SETSCREW

CONDUIT

11-29. *An entrance cap with setscrew mounting and snap-on cap. (Appleton)*

COMPRESSION NUT

CONDUIT TO ENTRANCE CAP

REMOVABLE COVER

CONDUIT INTO BUILDING

11-30. *An entrance ell. (Appleton)*

COMPRESSION NUT

BRASS SCREWS

cap is fastened to the building instead of being supported by the conduit. The cable is secured to the building with clamps or straps. Fig. 11-27.

    • *Thinwall entrance caps and elbows* are like the threaded rigid conduit caps and elbows. However, they are made for thinwall instead of rigid conduit. They are used, as previously stated, to bring power into a building. The cables are brought from the overhead supply outdoors to the entrance cap anchored on the building. Wires from the inside of the building are then brought into an entrance elbow mounted on the exterior of the building. They are pulled through the ell, the conduit, and to the entrance cap. Connections are made at the entrance cap. Often a 90° elbow is needed where the conduit enters the service switch. Figs. 11-28 and 11-29.

    Another type of entrance ell is shown in Fig. 11-30. It mounts to EMT with compression rings and nuts on either end. A removable cover facilitates the pulling of wires from one end and feeding them through the other.

11-31. *A 90° elbow installed. (Appleton)*

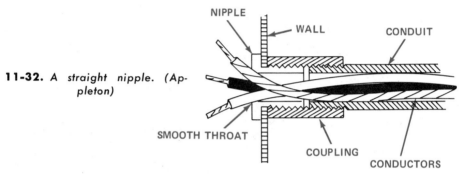

11-32. *A straight nipple. (Appleton)*

Another 90° elbow is shown in Fig. 11-31. It is available in a long or short size. You use it to connect a run of thinwall conduit to a metal enclosure with a knockout at a 90° angle to the conduit run. The EMT is held in the female end with a compression ring and nut. The male end is secured to the entrance with a tiger-grip locknut. The edges of all these fittings are chamfered to prevent possible damage to wire insulation and cable sheath.

● *Straight nipples or chase nipples* are used with conduit couplings to attach conduit to a box. Fig. 11-32. Place the nipple through the knockout from the inside of the box. Attach a coupling to the nipple, and the conduit to the coupling. To connect two boxes, place a chase nipple through each knockout from inside and connect with the conduit coupling. Chase nipples are available in sizes from $\frac{3}{8}$ inch to 5 inches.

● *Offset nipples* are designed to connect conduit to a box when the knockouts and conduit run do not line up. They are also used to connect two boxes side-by-side when their respective knockouts do not align. Fig. 11-33.

To connect conduit to a box using an offset nipple, secure the nipple to the box with a locknut. Connect the nipple to the conduit with a coupling. To connect two boxes, secure both ends with a locknut. Offset nipples are available in sizes from $\frac{1}{2}$ inch to $1\frac{1}{4}$ inches. They are made of malleable iron.

● *Conduit male enlargers* are vital fittings in conduit-to-box installations. They are used where the diameters of the conduit differ from those of the knockouts. They are made, in three sizes, of malleable iron. Plated over the iron is a cadmium finish. The

11-33. *Offset nipples. (Appleton)*

11-34. *Conduit male enlargers. (Appleton)*

fitting is installed by threading over the end of rigid conduit, then inserting it through the knockout. Fig. 11-34.

• *No-bolt fixture stems* are used with outlet boxes. They suspend fixtures from the ceiling without bolts. Male and female types are available. Both are threaded external-ly. The female stems also have $\frac{3}{8}$-inch internal threads. The shoulders have $\frac{1}{8}$-inch pro-trusions that match the knockouts on the outlet box. The fixture stem is placed through the base of the outlet box with the protrusions falling into place in the knock-outs. A locknut is then tightened onto the stem. This holds the fixture stem in place. Fig. 11-35. Both types are made with a cadmium finish to prevent rust and corrosion.

• *Tiger-grip locknuts* attach a piece of conduit or a connector into a knockout opening in a metal box. One inside locknut is used in connecting conduit to a box. An-other locknut is used on the outside of the box. The pressure exerted by the two lock-nuts locks the conduit firmly in place through the wall. Fig. 11-36.

11-35. *No-bolt fixture stem. (Appleton)*

11-36. *Tiger grip lock-nuts. (Appleton)*

11-37. *Snap-in blank. (Appleton)*

11-38. *Hex-head wedge adapter and hub body for rigid conduit. (Appleton)*

11-39. *Attaching a hub to the box and adapter. (Appleton)*

11-40. *Conduit bender.*

In tiger-grip locknuts there are non-slip notches around the perimeter of the locknut. These notches help prevent wrenches and screwdrivers from slipping. They also provide a perfect ground when tightened. You can be sure of the connection since their tilted edges bite into the metal wall of the box. The connection will not vibrate loose from excessive movement. Tiger-grip locknuts have been used where local codes demand an "approved" locknut.

• *Snap-in blanks for knockouts* are used to seal off knockouts made in error, or in rework in metal boxes. Fig. 11-37.

• *Rigid conduit hubs* consist of two parts: the body and the hex-head wedge adapter. The threaded shank of the hex-head wedge adapter is placed through the knockout from the inside of the box. The hub body is screwed onto the wedge adapter until the fitting is reasonably tight with the box. Conduit is then installed to the hub and the fitting tightened securely. Fig. 11-38. The hex-head design permits tightening from either outside or inside the enclosure. Fig. 11-39.

Often, when making an installation using threaded rigid conduit, junction boxes and device housings are used which do not provide suitable means for fastening the conduit to the enclosure. Knockouts, found in most enclosures, provide a location where rigid conduit hubs can be fastened. If there are no knockouts in the enclosure, a hole must be drilled.

The self-locking, hex-head wedge adapter and the hub body exert a continuous, uniform 360° pressure on the inside and outside surface of the box wall. This eliminates the need for locknuts. All hubs have a built-in, recessed neoprene gasket, and a flame-resistant insulated throat. The latter eliminates the need for an end bushing. It protects the wire insulation and cable sheath from damage due to vibration. The insulated throat leaves more wiring room within the enclosure.

Connectors are made with a zinc finish over steel. Larger sizes are of malleable iron with a cadmium finish.

Thinwall Conduit

Thinwall conduit is metal tubing. It is called EMT, electrical metallic tubing, by those who use it. This type of wiring is used with metal boxes only. There are connectors made for this type of wiring. They will be described and shown being attached in the following paragraphs.

Empty EMT should be mounted in place and connected to boxes and then the wire is pulled through it. Figs. 11-40 and 11-41.

11-41. *Conduit with wires pulled through. (Sears)*

Connector

Conduit

Strap

Coupling

11-42. One method of installing conduit for house wiring. (Sears)

[A] New style hanger for ceiling outlets. [B] Square box with mounting bracket. Use cover where code requires or space is needed for splices.

[C] Steel box support for mounting receptacle boxes between studs; also one or more switch boxes in any position.

Figure 11-42 shows one method of using conduit to wire a house. Note how the studs are notched to allow the conduit to be installed. Lengths shorter than 10 feet are cut with a hacksaw or tubing cutter. The ends are reamed so the cut ends are tapered and no jaggered edges are left to cut through the insulation on the conductors. Conduit should be supported with a strap every 6 feet on exposed runs. In some instances it would be better to support it at shorter intervals.

When connecting conduit to boxes, fit the threadless end of the conductor over the conduit, and insert the connector through the box knockout. Then tighten the locknut.

After conduit and boxes are installed, pull the wires through the conduit into the boxes. Allow 8 inches of insulated wire at each end of the box for connections. Use a white wire for neutral, and a black or red wire for the "hot" side.

In exposed work, conduit may be mounted on studs or rafters without additional protection. In concealed work, conduit must be supported.

Thinwall-conduit compression connectors and couplings are used with a special EMT (thinwall) conduit. Thinwall conduit connectors are employed in fastening (connecting) EMT to outlet boxes, switch boxes, panel and other metal enclosures. Fig. 11-43. Thinwall conduit *couplings* are electrical fittings used to attach (couple) the length of one conduit to another. They are used to extend the overall length of a piece of conduit. Conduit usually comes in 10-foot lengths.

Throat openings of the thinwall connectors and couplings are chamfered to eliminate burrs. That means the ends have been slightly reamed out to eliminate small metal bits that can damage wire being pulled through the conduit. All sizes of these connectors and couplings have hexagonal nuts, and bodies that must be held securely. A wrench is used to hold the nuts when tightening. The body of the fitting has a conduit stop that allows the EMT to enter evenly for uniform strength.

The ends of the connector, or coupling, house a compression ring. When the nut is tightened, the inward motion of the nut forces the open compression ring into a closed position around the conduit. This action locks the fitting onto the conduit.

● *Two-piece thinwall-conduit connectors* are used where thinwall conduit is to be connected to an outlet box, switch box, panel, or other metal enclosure. It has the advantage of increased wiring room inside the enclosure. Installation is fast with a single wrench. This means lower cost. Fig. 11-44.

COMPRESSION RING

NUT   BODY   LOCKNUT

11-43. *Compression connector. (Appleton)*

COMPRESSION RING

NUT

BODY   TAPERED SEAT

11-44. *Pressure cast coupling for EMT. (Appleton)*

11-45. *Two-piece conduit connector.* *(Appleton)*

11-46. *Thinwall connector in place on a box.* *(Appleton)*

This connector consists of a knurled, chamfered split-steel body with a hex nut. Fig. 11-45. To install, insert the body through the outlet box to the shoulder stop rim on the connector body. The conduit is then inserted through the nut into the body. Fig. 11-46. Tighten with a wrench placed over the hex nut. This draws the knurled chamfer of the body against the knockout hole of the box. At the same time, it compresses around the conduit. Thus, slippage is prevented and there is no need for a locknut.

Two-piece, thinwall-conduit connectors may be used to install conduit between two stationary enclosures. If used here, simply cut the conduit to correct length. Slip the hex nuts on both ends. Insert the conduit through the knockout in one box far enough so that it may be backed up through the knockout in the second box. (Backing up is shown in Fig. 11-47.) Place the connector bodies over the ends of the conduit from the inside of the box. Tighten with a single wrench. These connectors are concrete-tight and UL approved.

• *Indenter-type thinwall-conduit connectors and couplings* are used for attachment of thinwall conduit to metal enclosures. They are designed for joining two sections of thinwall. They provide a permanent, rigid connection or coupling. Install them with an indenter tool. Figs. 11-48, 11-49, and 11-50. All indenter fittings feature chamfered edges to prevent damage to cable sheath and wire insulation. Some have insulated throats.

Install by inserting the EMT into the fitting until it rests against the conduit stop. This allows the conduit to enter only to the halfway point. Here it is held in check. This provides the strongest possible connection. With the EMT inserted, the indenter-tool jaws are placed around the fitting and squeezed tightly. Fig. 11-51. The prongs make deep indentations in the fitting and conduit. The tool is then rotated 90° and another set of indentations is made.

Rotation and indentation are continued until a total of four have been made. The process is repeated for completion of coupling installation. Fig. 11-52. Indenter-type

EMT BACKED UP
INTO OUTLET BOX

HEXAGONAL NUTS

CONNECTOR BODY

11-47. *Two-piece connectors used between two boxes. (Appleton)*

THINWALL

CONNECTOR BODY

LOCKNUT

BOX WALL

INDENTER TOOL

11-48. *An indenter coupling. (Appleton)*

"One-Squeeze"
Indenter Tool

B-M

Steel Indenter
Connector

B-M

Steel Indenter
Connector
3/8" Offset

B-M

Steel Indenter
Coupling

B-M

Steel Insulated
Indenter Connector

11-49. *Indenter connectors and an indenter tool.*

CONDUIT

LOCKNUT

INDENTER MARKS

INSULATED
THROAT

11-50. *Indenter connector installed. (Appleton)*

INDENTER TOOL

INDENTER COUPLING

INDENTER MARKS

CONDUIT

11-51. *Indenter coupling. (Appleton)*

279

connectors have a tiger-grip locknut. The locknut is tightened on the inside of the enclosure for a slip-proof bond. Fig. 11-50. They are, of course, plated with cadmium to prevent rust or corrosion.

• *Setscrew thinwall-conduit connectors and couplings* are used where EMT is to be fastened (connected) to a switch, outlet, or panel box. Setscrew couplings are applied where one run of thinwall conduit is to be attached (coupled) to another. These connectors and couplings are designed for use on straight runs of conduit; that is, where two ends of the thinwall meet and line up properly. They may also be used where the two ends of the conduit line up with the knockouts of a box. Figures 11-53 and 11-54 show a setscrew connector.

To install, simply loosen the setscrews on the fitting. Then insert the thinwall conduit until it falls against the built-in stop collar inside the fitting. This stop collar assures even holding pressure for EMT. Also, it provides a smooth surface over which

11-52. *Indenter coupling cutaway. (Appleton)*

11-53. *Setscrew connector. (Appleton)*

11-54. *Installed setscrew connector. (Appleton)*

11-55. *Cutaway view of a coupling, setscrew type. (Appleton)*

ENTRANCE
FLEXIBLE
CONDUIT

ENTRANCE
THINWALL
CONDUIT

COMPRESSION
RING

TIGHTENING
SCREW

BODY   NUT

FLEXIBLE
CONDUIT

THINWALL
CONDUIT

TIGHTENING SCREW

11-56. *Thinwall to flexible-steel conduit fitting. (Appleton)*

LOCK LIP

MALLEABLE IRON

OVAL
SCREW HOLE

STEEL

STEEL CLAMPS

RIGID OR
THINWALL
CONDUIT

SCREW OR BOLT

KNOCK-OUT   CONDUIT

CLAMP BACK

KNOCK-OUT LINES UP CONDUIT
NO OFFSET BEND REQUIRED.

11-57. *Conduit clamp, one-screw type. (Appleton)*

wires may be pulled easily without damage to the insulation. All edges are chamfered for further protection.

After inserting the conduit, tighten the setscrews against the outer wall of the conduit. Deep-slotted, staked setscrews thread firmly into the embossed surface of the body of the fitting. Repeat the procedure to install the conduit in the other side of the coupling. To complete the connector installation, insert the conduit and the connector in the box knockout. Tighten the locknut to the interior of the box to prevent slippage. Fig. 11-55.

The combination flexible-steel conduit to EMT coupling is manufactured in $\frac{1}{2}$-, $\frac{3}{4}$-, and 1-inch sizes. One end has a split body with a tightening screw for flexible-steel conduit to a thinwall fitting. Place the end with the split body over the flexible conduit. Tighten the screw. Fig. 11-56. Then insert the thinwall conduit in the compression ring end and tighten the nut for a rigid bond.

A *conduit stop* inside the body of the fitting assures an even strength of bond. All edges are chamfered to prevent damage to cable sheath and wire insulation. Heavy-gage steel fittings are precision machined. They have a cadmium finish to resist rust and corrosion. At the open end of the clamp is a lip that locks conduit into place. An oval screw hole allows adjustment after installation. Sizes are stamped on the back of the clamp for easy identification. Fig. 11-57.

● *Two-hole pipe clamps,* or straps, are flush-mounted supports for fastening conduit to a wall or ceiling. They are light in weight, yet durable and versatile. On both sides of the arch of the two-hole pipe strap are protruding punch marks that serve to lock the conduit or cable in place. Identification is easy. The size is marked on the back of each strap. Fig. 11-58.

• *Adapters, clamps, and straps for thinwall conduit* are made for a number of jobs. The adapters are used for converted threaded, rigid-conduit fittings for use with thinwall conduit. Fig. 11-59.

The adapter is a compression-type fitting with external threads. It has a split body. Thinwall conduit is placed inside the adapter and the assembly tightened into the threaded fitting. As you tighten the adapter, it is forced into a closed position. This closed position around the EMT forms a rigid bond.

11-58. *Conduit clamp, two-screw type. (Appleton)*

11-59. *Installation of threaded rigid conduit to thinwall conduit adapters. (Appleton)*

Conduit clamps and straps are used to fasten conduit to walls, floors, and ceilings.

• *One-screw steel clamps* are used in flush-mounting thinwall. They are available in ½-inch to 2-inch sizes. An oval-shaped screw hole allows for adjustment after installation. The size is stamped on the back of the clamp. Fig. 11-60.

• *Two-hole pipe straps* are flush-mounting supports for thinwall. They are inexpensive and are available in sizes up to 1½ inch. Protruding marks on both sides (raised above the surface) of the strap serve to lock the conduit in place. The size is stamped on the back of the strap. Fig. 11-61.

Note that there is a distinct difference between clamp diameters for rigid conduit and those for thinwall conduit.

Bushings

• *End bushings* are used where threaded rigid conduit enters a metal enclosure. These enclosures may be boxes for controls, panel boards, or fuse boxes. A method to protect the wire from abrasion must be provided. This is necessary because the end of the conduit is very rough and is therefore capable of cutting the conductors. This cutting of the conductors causes a dead short or other damage. For this purpose, end bushings are designed to thread onto the end of the conduit so that the teeth will bear against the outlet box in which the conduit is fastened.

The end of the conduit is then slipped through the knockout. Fig. 11-62. There is a raised, rounded surface in the bushing over which the wire slides. The wire does not touch the conduit at this end. Fig. 11-63. After the bushing is installed, tighten the locknut on the outside of the box. The teeth of the locknut dig into the metal box, making a continuous ground.

11-60. Mounting of steel EMT clamp. (Appleton)

11-61. A two-hole pipe strap installed. (Appleton)

There are several types of end bushings available. For example, malleable-iron bushings secure the conduit in the same manner as a locknut. Plastic bushings protect the end of the conduit, but require the addition of locknuts for holding.

CONDUITS AND THE CODE

EMT is a general-purpose raceway. It is similar to rigid conduit and IMC (intermediate metal conduit). It is between rigid conduit and EMT in wall thickness. It does use the same threading method and standard fittings as rigid metal conduit. Generally speaking, it also fits the bill for rigid conduit wherever it is called for. The only difference is that IMC is slightly lighter, since its wall thickness is less than that of rigid conduit. It requires about 25% less steel than rigid conduit. It performs about as well as rigid in most places. It is better, of course, than rigid aluminum and EMT in most locations. It is less expensive than rigid, but more expensive than EMT. Its out-

11-62. Locknut and bushing properly installed. (Appleton)

11-63. Thermoplastic bushings. (Appleton)

side diameter is the same as that of rigid conduit. This larger inside diameter makes it easier to pull wires through. However, the NEC does not allow for more wires to be included because of its greater inside diameter.

Comparison of Rigid Conduit and IMC

The $\frac{3}{4}$ inch trade sizes of both rigid conduit and IMC are shown in Table 11-A.

CLAMPING OF IMC. A clamp or other type of support must be placed within 3 feet of any box, cabinet, or conduit termination. There shall be not more than 10 feet between any pair of clamps for any IMC at any angle.

NUMBER OF CONDUCTORS. The number of wires allowed in rigid, IMC, or EMT is the same for conduit $\frac{1}{2}$ inch and larger.

IMC cannot be used when buried in cinders. Cinders have sulfur content. If there is moisture, then the cinders present a sulfuric acid. This acid corrodes the metal. The IMC must be buried at least 18 inches below the cinder fill. However, it can be buried in the cinders if encased in 2 inches of concrete.

A run of conduit cannot have more than 360° of bends from the point of outlet to outlet. This includes the bends at the outlets. For example, if you have a 45° angle at the outlet box and another 45° so the conduit fits against a flat surface this counts as 90°. If the same bends are used at the other box, then another 90° is used. That leaves only 180° of bends for the entire length of the run from box to box or panel to box.

Aluminum conduit definitely requires additional corrosion proofing if it is buried in concrete. Galvanized steel conduit, rigid conduit, and galvanized IMC installed in concrete do not require supplementary corrosion treatment.

RIGID NONMETALLIC CONDUIT. Materials used for this type of conduit are fiber, asbestos, cement, soapstone, rigid polyvinyl chloride, and high-density polyethylene for underground use. Rigid polyvinyl chloride is used aboveground.

Plastic conduit is a better name for rigid nonmetallic conduit. It is permitted to be used where the voltage does not exceed 600 volts. It can be used in walls, floors, and ceilings. This type of conduit lends itself for use in cinder-fill situations. However, if the voltage is over 600 volts, it must be encased in 2 inches of concrete.

Plastic conduit must be used where the temperature does not exceed its rating. It should not be used where there is the possibility of physical damage. It should not be used to support fixtures or other equipment. Support of the conduit must be as shown in Table 11-B. In addition, it should be noted that there shall be a support within 4 feet of each box, cabinet, or other conduit termination.

Polyvinyl chloride conduit and high-density polyethylene conduit are made and shipped in 10-foot lengths. Conduit bodies must have a cross-sectional area at least twice that of the largest conduit to which they are connected.

The number of conductors in conduit is limited by the National Electrical Code. For new work or rewiring, the total cross-sectional area of the conductors must not be over 40% of the internal cross-sectional area of the conduit. For three or more conductors, the sum of their areas must not exceed 40% of the conduit area. For details on the fill of conduit check the latest issue of the National Electrical Code.

Aboveground Use

PVC (polyvinyl chloride) conduit is the only rigid *nonmetallic* conduit that may be used aboveground. There are two schedules of rigid nonmetallic conduit — Schedule

Table 11-A. Comparison of Rigid Conduit and Intermediate Metal Conduit

Conduit	Outside diameter	Inside diameter	Wall thickness
Rigid steel	1.050″	0.824″	0.113″
IMC	1.050″	0.908″	0.071″

Table 11-B. Support for Rigid Nonmetallic Conduit

Conduit size (in inches)	Maximum spacing between supports (in feet)	
	Conductors rated 60° C [140° F.] and below	Conductors rated more than 60° C [140° F.]
½ to ¾	4	2
1 to 2	5	2½
2½ to 3	6	3
3½ to 5	7	3½
6	8	4

80 and Schedule 40. Schedule 80 has an inside diameter that is less than Schedule 40. For conductor fill to 40% of the cross-sectional area, the wire fill capacity is marked on the conduit surface.

This plastic PVC conduit can be used aboveground for over 600 volts if it is encased in at least 2 inches of concrete. It can be buried if it meets the depth requirements of the Code. This type of conduit can be obtained with a diameter of ½ inch to 6 inches. One coupling is furnished with the 10-foot length of conduit. This type of conduit can be used where there are corrosive agents, such as in industry. It is also useable where there are vapors, caustic mist, pickling acids, and plating baths. In home use you would probably use it to mount spotlights outdoors on poles or along the top edges of the house. This type of conduit is cemented together at junctions or couplings with cements made for the specific type of plastic.

There are factory-made elbows and fittings for the PVC and other types of plastic or nonmetallic conduit. Fig. 11-64.

Hotbox benders are available for bending the plastic conduit. It simply heats the plastic so it can be bent to the desired angle.

Take a look at Fig. 11-65. This is the standard mounting for a box if supported by two threaded conduits stubbed up out of concrete or the ground. However, it cannot be used for nonmetallic conduit.

These limitations must be kept in mind for residential wiring. Sometimes it is desirable to have a patio plug or flowerbed plug available for operating power equipment. The maximum height of 18 inches helps the pipe or conduit withstand blows that may be accidentally applied.

11-64. *Factory-made fittings for PVC conduit. (A) bell ends and caps, (B) reducer, (C) female adapter, (D) bell ends and caps, (E) coupling, (F) terminal adapter, (G) adapter, (H) PVC cement (quart size), (I) PVC cement (pint size), (J) expansion joint. The access fittings on the right are identified by type.*

11-65. *Mounting boxes and lamps on conduit.*

FLEXIBLE STEEL CONDUIT

U.L. & C.S.A. LABELED FLEXIBLE
STEEL CONDUIT

ALUMINUM FLEXIBLE CONDUIT

FLEXIBLE LIQUID-TIGHT CONDUIT

11-66. *Flexible steel, aluminum, and liquid-tight conduit.*

FLEXIBLE CONDUIT

This type of conduit is known as Greenfield or flex. Fig. 11-66. It comes in coils no longer than 250 feet. However, it can be bought in coils of 50 and 100 feet. The $\frac{5}{16}$-inch size as well as the $\frac{3}{8}$-inch size comes in 250-foot coils. The $\frac{1}{2}$-inch size is available in 100-foot coils and the other three sizes ($\frac{3}{4}$-, 1-, and $1\frac{1}{4}$-inch) come in rolls of no more than 50 feet.

It may be used in runs no longer than 6 feet with no requirement for a separate ground wire. Flex can be obtained in either aluminum or steel. It has been judged to be suitable for a ground only up to 6 feet in length. When you use the aluminum flex, you must be careful not to tighten the connectors with setscrews too tightly. This will damage the flex. On flexible conduit made of aluminum you will find the AL stamped every foot. This identifies it as aluminum. Flexible conduit should not be used in wet locations. However, it is permitted to be used if the conductors inside have a W designation. For example, THW or XHHW would be types of conductors useable inside the flex in a wet location since they have a W designation.

If the W-type wire is used, it is also obvious that you should use watertight connectors. They will prevent water from running into equipment at the terminating end.

Supports

Flexible conduit requires support at much closer intervals than rigid conduit. Supports should be every $4\frac{1}{2}$ feet and at least within 12 inches of a box or terminal. The bends made in the flex should be held rigidly so it will not deform when wires are pulled through it. There is an exception to this rule. A 5-foot length of ⅜-inch flex can be used with two No. 14 AF wires with no grounding conductor and without support to supply power to a recessed incandescent lamp. Such a lamp is shown in Fig. 11-67.

Sealtite Flexible Cable

There are some uses for this type of waterproof flexible cable around the house. It can be used in wet locations and comes in sizes of ⅜ inch to 4 inches. It does require a few special connectors for it to retain its waterproof condition. These are shown in Fig. 11-68.

• *STCN connectors* are made for type CN nonmetallic sealtite conduit. STCN means *seal-tight connector, nonmetallic.* CN stands for *conduit, nonmetallic.* STCN connectors are made for use with Anaconda® Type CN sealtite conduit. Fig. 11-69. (*Sealtite* is the usual spelling for *seal-tight* conduit.) This type of conduit (CN) differs from standard metallic sealtite conduit in that it does not utilize a flexible metallic interior. Instead, CN conduit has a smooth plastic inner core, covered with a bonded reinforcing nylon cord. It also has a rugged plastic outer jacket for covering. The absence of any metallic elements in the conduit eliminates the problem of attack by corrosive atmospheres.

The STCN connector also has the advantage of making the conduit extra flexible. It provides an extremely smooth interior for protection of conductors. It is easily identified by its bright orange color. STCN connectors form a liquid-tight and vapor-tight connection of CN conduit to a circuit breaker, switch, junction box, and similar enclosures, as well as to boxes and threaded hubs.

The STCN connector has four simple, easy-to-install parts: a nylon ferrule, a steel body, a compression nut, and a tiger-grip locknut. Fig. 11-70. The steel body of the connector has a built-in, protective plastic-insulated throat. It also has a neoprene "O" ring.

To install the connector:

1. Cut the CN connector to the desired length, being sure to make a very straight cut.

2. Feed the compression nut onto the conduit.

3. Press the threaded end of the nylon ferrule as far as possible into the end of the conduit with the palm of the hand. Use the connector body as a wrench to tighten the

11-67. *Using flexible conduit for connecting an incandescent lamp fixture to a conduit or a cable power supply.*

Conduit Or Cable Circuit

Outlet Box

Flexible Conduit

Ceiling

Incandescent Fixture

11-68. *Flexible conduit and connectors. (Appleton)*

FLEXIBLE LIQUID-TIGHT CONDUIT

FERRULE

CONNECTOR BODY

NUT

FERRULE THREADED INTO CONDUIT

ASSEMBLY AFTER COMPRESSION SHOWING POSITIVE GROUND.

OUTER JACKET   SMOOTH INNER CORE

CN CONDUIT

BONDED REINFORCEMENT

11-69. *Type CN sealtite conduit. (Appleton)*

ferrule flange snugly to the end of the conduit. (The interlocking hexagonal-nut type design of the ferrule and body make this possible.)

4. Bring the compression nut up and tighten to a positive step on the shoulder of the connector body.

5. Place the end of the body through the knockout and secure with the locknut provided. Fig. 11-71. Check Fig. 11-72 for installation of conduit into a threaded unilet hub.

CN CONDUIT

COMPRESSION NUT

NYLON FERRULE

CONNECTOR BODY

LOCKNUT

11-70. *Installing a type CN nonmetallic seal-tite conduit conductor. (Appleton)*

11-71. *Attaching CN conduit to a wall box. (Appleton)*

11-72. *Attaching the CN conduit to a unilet hub. (Appleton)*

QUESTIONS

1. Why is conduit usually not used for wiring houses?
2. What do you use to cut the metal covering on armored cable?
3. How are BX connectors different from Romex?
4. What sizes of armored cable are available for residential work?
5. What does ACT mean on BX cable?
6. What does an L mean when labeled on a piece of BX?
7. What is induction?
8. How do you secure BX to studs or supports?
9. How much clearance must be maintained between a service drop and a window on the house it is serving?
10. What is an entrance elbow?
11. What is a pulling elbow?
12. What is the difference between thin-wall and rigid conduit?
13. What is an offset nipple?
14. What is a locknut?
15. What is the standard length of thin-wall conduit?
16. What does the EMT stamped on a piece of conduit mean?
17. What is the purpose of an end bushing?
18. What is IMC?
19. How many bends can you have in a piece of conduit?
20. Can polyvinyl chloride conduit be used aboveground?
21. What is another name for flexible conduit?
22. Can flexible conduit be used in wet locations?
23. What does W on a piece of wire's insulation mean?
24. Where can sealtite flexible cable be used?

Remodeling Wiring

Whenever remodeling is considered, electrical requirements must be considered and given top priority. Each remodeling job is different. The electrician must be versatile and ingenious. Situations will arise that will not be covered in any book or any Code. It should be remembered, though, that good wiring practices must be followed. The Code should be followed. However, since the variations created by the physical environment will alter the conventional methods, it may be difficult to follow the Code. It may take some ingenious methods, but in most cases, it can be done with a little thought and planning.

Needless to say, remodeling will take longer than new house wiring with drywall still to be installed. Each situation involves planning and the checking out of the old construction techniques.

Remodeling wiring usually calls for "adding on" to an existing system. The old system should be updated at this point, if necessary. That is, of course, if the owner is willing to provide the funds for such rewiring.

Figure 12-1 shows how to replace the older type of push-button switch with a more modern toggle-type of on-off switch. Note that the face plate must also be changed.

Cover plate   Switch

12-1. *Replacing an old push-button with a toggle switch. (Sears)*

Green screw

Cover plate

Switch

Uninsulated ground wire

If no green screw is located on the switch, connect wire as shown here.

Wall of box

Grounding wire to receptacle

12-2. *A fuse and circuit tester.* *(Sears)*

Before replacing the switch, a simple circuit tester should be used to determine if the circuit is live. A device such as that shown in Fig. 12-2 can be used. If the device glows when plugged into the outlet or across the proper terminals, it means the circuit is *live,* or *hot.*

CIRCUIT TESTING

Before replacing or adding onto an existing system, test the circuits to see if the wiring has been done correctly. This can be done in a number of ways. Figure 12-3 shows a circuit tester designed to locate faults in wiring and to help check for continuity in existing outlet circuits. The circuit tester can be used to test circuits with outlets such as those shown in Fig. 12-4. By using an adapter, it is possible to check other types of circuits without the added neutral. This tester has a simple visual set of neon color-coded indicating lights to identify various fault conditions in electrical circuits. The arrangement of lighted and/or unlighted indicating lamps will immediately show the circuit condition. Fig. 12-5.

ADDING ANOTHER OUTLET

Switch and outlet boxes for additional outlets or switches must be located between studs. This will allow room for inserting the box and box supports. Decide where the

12-3. *A receptacle polarity circuit tester.* *(Bryant)*

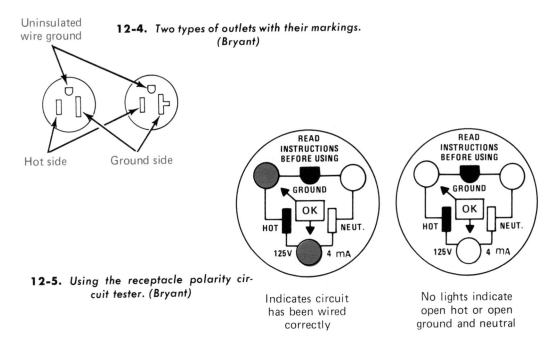

Uninsulated wire ground

12-4. *Two types of outlets with their markings. (Bryant)*

Hot side   Ground side

12-5. *Using the receptacle polarity circuit tester. (Bryant)*

READ INSTRUCTIONS BEFORE USING
GROUND
OK
HOT   NEUT.
125V   4 mA

READ INSTRUCTIONS BEFORE USING
GROUND
OK
HOT   NEUT.
125V   4 mA

Indicates circuit has been wired correctly

No lights indicate open hot or open ground and neutral

box is to be located for the switch or the outlet. Mark the spot. Switches are located from 48 to 54 inches above the floor. They should be on the same level as existing switches. Convenience outlets are located about 12 to 18 inches above the floor. In the kitchen or dining area, the location may be different since they are needed for toasters and other appliances. That means they will be slightly above the dining table or kitchen working space.

Switches should be placed at the opening side of the door. Use the 2½-inch deep box when possible.

Follow the steps shown in Fig. 12-6 for installing boxes. Use the sounding method — thumping on the wall — until you locate the studs. Some devices are available with magnets that point toward nail heads in the drywall or sheetrock, thus locating the stud.

Figure 12-6 shows the method used to install an additional box in old lath and plaster. Placing the box into a drywall installation is much simpler. Be sure to place a floor covering down to catch falling plaster or dust.

When cutting the hole in the wall with the hacksaw blade, hold the board or plaster with your hand to prevent cracking. After completing the eight steps shown in Fig. 12-6, look at Fig. 12-7A. This shows one type of box to be used in this type of add-on wiring. This box has "grip-tight" brackets to hold the box in place. Figure 12-7B shows another type of box being used with metal box supports. Insert the supports on each side of the box. Work the supports up and down until they fit firmly against the inside surface of the wall. Bend the projecting ears so that they fit around the box.

12-6. Steps in installing a box in an old building. (Sears)

Use hacksaw

Install locknut

Use template

Pull in connector

Notch out plaster

Insert box

Sound for studs

Cut laths

When it is possible to fasten the box in place, look at Fig. 12-8. If either outlet is on an "interior" wall, you can usually drill straight upward to the partition between the walls. Secure the cable properly with staples as called for in your local code. Figure 12-9 shows how to run the cable across the basement.

WALL SWITCH ADDED TO CONTROL OUTLET

Most modern homes do not have overhead lamps in the bedrooms. They rely upon wall outlets to provide power for floor or table lamps. If a switch is needed to control a lamp, it is possible to add it without going through the basement. Fig. 12-10. This will be handy if you are in an upstairs bedroom and there is a finished ceiling below.

First, cut a hole in the wall. Remove the baseboard and cut holes B and C. Notch a channel in the plaster where the baseboard was originally. Remove the knockout in the switch box (A). Attach a cable connector. Fish the cable into the knockout in A, from B. Then run it through hole C to box D. Connect as shown. *Do not forget to attach*

12-7A. *Using a box with a "grip-tight" bracket to hold the box in place in the addition of a switch or outlet to existing wiring. (Sears)*

12-7B. *Method of installing a box with metal box supports. Insert the supports on each side of the box. Work the supports up and down until they fit firmly against the inside surface of the wall. Bend the projecting ears so that they fit around the box. (Sears)*

All fish tapes (or snakes) must have a hook for fishing thru conduit, ceilings or partitions. This hook is made by heating end of fish tape on open flame until wire becomes red hot and bending end of fish wire with pliers to correct size hook.

12-8. *Fishing a wire with two fish wires. (Sears)*

the ground wire. Attach it to the switch and the plug if they have a screw for doing so. If not, attach it to the box with a screw or slip-on connector. Fig. 12-11 and Fig. 12-12. Replace the baseboard.

ADDING AN OUTLET ALONG THE WALL

Adding an outlet along the wall involves the procedure explained above. Figure 12-13 shows how an additional outlet is added to the wall. Extend a piece of wire from box E to box F. Place the cable in the groove cut for it in the plaster and replace the baseboard. Connect the wiring according to the color code. Connect black to black

12-9. *Wiring two outlets — extending one to another wall by means of the basement floor. Drilling through the floor is similar to the operation shown in Fig. 12-8. (Sears)*

12-10. Removing the plaster between the two points and under the wallboard makes it easier to install a switch and outlet box. (Sears)

12-11. Romex connector attached to a box. Note the ground clip for the uninsulated ground wire. (Sears)

Uninsulated grounding wire

Switch box

A

12-12. (A) Grounding the wire to a screw on the side of the metal box. (B) Slater Electric's method of eliminating the need for running a grounding wire from the green terminal of the switch or duplex outlet. (Slater)

Special straight wire spring

Squeezed, four-lobed mounting screw

B

12-13. A wall to wall baseboard outlet extension. (Sears)

12-14. *Wiring another box back-to-back with an existing outlet in another room. (Sears)*

and white to white. Then run them to the new box. Make sure the black wire is connected to the brass-colored screw and the white wire to the silver-colored screw. Attach the ground wire (uninsulated) as shown in Fig. 12-11 or Fig. 12-12.

If you wish to place outlets in another room back-to-back with an existing outlet, look at Fig. 12-14. Use matching holes. Wire with conduit or threaded nipple with locknuts. Ground wires can be attached to the boxes with screws or clips.

RUNNING A WIRE AROUND A DOOR

Adding a new wall outlet may be difficult in some instances. Fig. 12-15. Remove the baseboard and the trim around the door. Notch the wall and spacers between the frame and door jamb. If the outlet is to be placed beyond the first upright, use an extension bit to drill past the additional uprights. Wire the boxes and replace the baseboard and the door trim.

In some cases, headers are not always found in partitions. Fig. 12-16. They can be bypassed by using the method shown in 12-16A. They can also be notched, as shown

12-15. *Running a wire around a door frame from the old outlet to a new one. (Sears)*

12-16. *A. Running a cable around a header. B. Running a cable through a header. (Sears)*

A   B

in Fig. 12-16B. The hole can be patched and repainted or wallpaper can be replaced carefully so as not to show the spot.

RUNNING A WIRE THROUGH THE FLOOR

In some cases, it is necessary to run a wire from one level of the house to another. As shown in Fig. 12-17, it is possible to add an outlet that is switched with the lamp. It is also possible to use an upstairs outlet and add a ceiling lamp in the basement or floor below. Select an outlet site and cut an opening through the lath and plaster.

If the outlet is on an outer wall, bore a diagonal hole from the basement. Use a long shank bit—if using a hand brace. Use a flat-type drill bit if using an electric hand drill. If the outlet is on the interior wall, bore through the floor to the partition between the

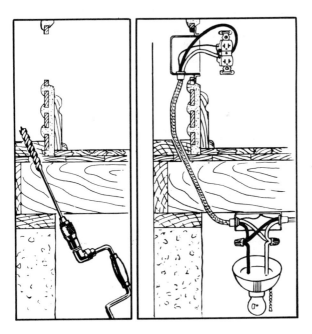

12-17. *Running a wire through the floor from the basement to the upper floor. (Sears)*

If you can get into attic or upper room, simply remove the upstairs baseboard. Then drill diagonal hole downward as shown.

Drill diagonal hole upward from opposite room. Then drill horizontally till holes meet. This method requires patching plaster.

Push 12-foot fish wire, hooked at two ends, through hole on 2nd floor. Pull one end out of switch outlet on 1st floor.

Next, push 20-25 foot fish wire, hooked at both ends thru ceiling outlet (arrows). Now fish until you touch the first wire.

Then withdraw either wire (arrows) until it hooks the other wire; then withdraw second wire until both hooks hook together.

Lastly, pull shorter wire thru switch outlet. When hook from long wire appears, attach cable and pull thru wall and ceiling.

12-18. *How to fish a wire. (Sears)*

walls. You do have to be careful to locate the wall and the partition 2 × 4. Use a piece of fish wire and push the cable up from the basement hole. Attach it to the wires and pull them through the outlet opening.

Attach the cable to the boxes as shown. Be sure to follow the correct color code for the outlet's brass and silver screws. Attach the extra uninsulated wires to the boxes with screws or clips.

HOW TO FISH A WIRE

Figure 12-18 illustrates the method of using a fish wire to pull wire through holes drilled through the floor and wall. The fish wire comes in handy when a thermostat is relocated or a humidity control or an air conditioning thermostat is added.

If you get into an attic or upper room, simply remove the upstairs baseboard. Drill a diagonal hole downward as shown in Fig. 12-18(A).

Drill the diagonal hole upward from the opposite hole. Then drill horizontally till the holes meet. This method requires patching the plaster. Fig. 12-18(B).

Push the 12-foot fish wire, hooked at two ends, through the hole on the second floor. Pull one end out at the switch outlet on the first floor. Fig. 12-18(C).

Push 20 to 25 feet of fish wire, hooked at both ends, through the ceiling outlet (note the arrows). Now fish until you touch the first wire. (Fig. 12-18(D).

Withdraw either wire (see the arrows) until it hooks the other wire. Then withdraw the second wire until both hooks come together. Fig. 12-18(E).

Pull the shorter wire through the switch outlet. When the hook from the long wire appears, attach the cable and pull it through the wall and ceiling. Fig. 12-18(F).

ATTIC INSTALLATION

The attic in older homes has floorboards. These can be carefully lifted and then re- placed after the rewiring has been completed. Figure 12-19 shows how the floorboards

12-19. *Lifting the floor of an attic to get at the ceiling for purposes of wiring an outlet or lamp. (Sears)*

12-20. *Lift the floorboards and place the wires in a notch or drill through the rafters or joists and feed the wire through the holes. Be careful in reinstalling the floorboards to make sure the nails do not penetrate the cable. (Sears)*

$1\frac{7}{32}$ in.

TOP

$\frac{1}{4}$in.

12-21. *Installing a secure type of outlet for outdoor use. (Pass & Seymour)*

can be lifted. The joists can be notched to hold the cable. A hole can be bored through obstructions with an electrician's bit. If the attic is inaccessible, but the cable can be run parallel with the floor beams or joists, connect the ceiling outlets with wall switches. Do this by drilling as shown in Fig. 12-20. Draw the cable through the opening using fish wire. Where the cable runs across beams, the floor boards must be lifted. *When reinstalling the boards, be careful not to drive nails through the cable.*

INSTALLING A SECURE BOX

In some instances, it is necessary to add a secure outside outlet. Keep in mind that outside outlets should have a GFI (ground fault interrupter). Providing the box and outlet with a key ensures that the plug is not used unless the person with the key is available.

Figure 12-21 shows how the adapter is fastened to a standard 4-inch or $4\frac{11}{16}$-inch box. The side marked "top" is placed at the top of the box. Run the conduit to the box as usual in the case of a brick wall. It is possible to use Romex in regular frame construction. Set the box in the wall so the front edge of the adapter will be flush, or recessed not more than $\frac{1}{4}$ inch from the finished wall.

Install the finish frame and cover after applying caulking compound to the back of the frame to provide a weatherproof seal. Fig. 12-22. Elongated slots in the frame allow for alignment.

Wire the device used — in this case the duplex outlet — and install it in the box. Note that the device must have a special mounting plate. Fig. 12-23. The 30- and 50-ampere receptacles should be installed with the ground contacts at the top to facilitate the use of angle cord sets.

Other Outside Switches and Receptacles

The switches and receptacles used outside must be weatherproof. Weatherproof devices must be installed where motors and high-current farm equipment are plugged into the circuit. These switches and receptacles vary in size. Fig. 12-24. Keep in mind that a motor with a rating of more than $\frac{1}{2}$ horsepower [373 watts] must have 240 volts to supply it properly. This will require three-wire installation and types of plugs different from those used on 120-volt. One-horsepower [746-watt] motors draw about 6.5 amperes on a 115-volt line. This means there will be little current for other appliances or lights on the 15-ampere line. Therefore, it is best to use the three-wire, 240-volt line for motors. Remember, there is some voltage drop along the wire. This is especially true if the distance from the circuit breaker box is more than 25 feet.

When wiring the outside boxes, make sure the black wire goes to the brass-colored screw on the receptacle. If a red wire is used, it goes to the brass-colored screw since it, too, is a hot line. The white goes to the silver-colored screw on a receptacle.

12-22. *Place the finish frame over the square box. (Pass & Seymour)*

12-23. *Install the GFI or outlet. (Pass & Seymour)*

| 10-ampere toggle switch | 15-ampere covered receptacles. Choice of single or duplex types. | Weatherproof fused main switch for mounting outdoors on pole or building wall |

12-24. *Weatherproof switches and outlets for outdoor use. (Sears)*

UNDERGROUND WIRING

You may want to add a light or garage door opener to a garage that is located some distance from the house. This can be done by placing the wiring underground. Fig. 12-25.

Check your local code to see which type of wire is approved for wiring underground. You may want to use an installation similar to that shown in Fig. 12-25. Here, conduit is used with waterproof fittings. If you do not care for the extra expense of this type of wiring you can substitute (if your code allows) the underground cable which resembles Romex.

By using the plastic-coated wiring, it is possible to link houses and outbuildings easily. Fig. 12-26. Farms can use this type of wiring easily and bury it. By burying the wire, it is protected from weather, vehicle traffic, and other hazards. In some localities, you are required by the local code to use only conduit. Check to make sure before installing plastic cable. Plastic cable can also be used to run through concrete or masonry. Today's wire is treated so that rodents will not chew on it when it is buried.

The National Electrical Code allows the use of type UF plastic wire for burying underground. It comes in No. 14 through 4/0 for use as branch circuit and underground feeders. It should be buried no less than 12 inches if it is used for residential circuits with less than 300 volts and no more than 30 amperes.

Wet and Damp Locations

In most instances, underground wiring will terminate in an outlet. This outlet may be a receptacle, switch, or fixture. Installation of a receptacle in a damp location is

12-25. Wiring underground with weatherproof conduit. (Sears)

Entrance switch or 3-way light switch inside house

3-way light switch or branch entrance switch in outbuilding

12-26. Using plastic wiring underground. (Sears)

governed by the Code. An open porch or a screened-in porch is a damp location. A conventional receptacle cannot be used in this location. The receptacle must have a cover plate that is weatherproof. The UL lists plates according to use, such as "weatherproof." The cover plate may have a screw-on cap such as that shown in Fig. 12-27. Figure 12-28 shows the side-opening type of covers for a receptacle. If the doors stay open, they are suitable for use in damp areas, such as porches. If such covers are self-closing, they can be used outside to supply portable tools or equipment.

A receptacle used in a wet area must keep the receptacle waterproof at all times when the plug is inserted and when it is removed.

Figure 12-29 shows a swing-down, self-closing canopy. It keeps the area shielded from rain when the receptacle is in use or plugged in. A duplex cover is also available for outdoor use. The NEC also requires an elevation of outdoor receptacles to prevent accumulation of water.

COVERS AND CANOPIES

The NEC requires (and it is commonsense to do so) that every outlet box be covered by a cover plate, a fixture canopy, or a faceplate. The openings for a receptacle, snap switch, or other device installed in the box must be covered.

One part of the Code requires all metal faceplates to be grounded. Because metal faceplates are *exposed* conductive parts, they must meet Code requirements. This means that ungrounded metal faceplates shall not be installed within 8 feet vertically

12-27. A chain-held screw-cap cover for damp locations.

12-28. Side-opening covers. If the doors are self-closing, such a cover assembly could be used outside. It could be used to supply portable tools if the equipment is not left connected.

Grounding terminal

Bare wire

Grounding conductor

Non-metallic box

Cable ⟶

12-29. *This swing-down canopy is self-closing. The duplex cover is also self-closing.*

12-30. *This type of switch must be used with metal switch plates mounted on nonmetallic boxes.*

or 5 feet horizontally of ground or grounded objects. This refers to laundry tubs, bathtubs, showers, plumbing fixtures, steam pipes, radiators, or other grounded surfaces that may be contacted by people.

Another part of the Code requires metal faceplates to be grounded in wet or damp locations, unless the faceplates are isolated from contact. Such locations can include bathrooms. It is a good idea not to install metal faceplates on switches or receptacles in bathrooms or kitchens. They may be hazardous if the home does not have the proper grounding system. If a nonmetallic box is used for the receptacle or switch, the grounding type of switch must be used. Fig. 12-30. **Note: In Fig. 12-30, the grounding conductor is not red, but uninsulated. Red is shown only for emphasis.** For any metal faceplate on nonmetallic boxes the grounding conductor from the feed cable must be attached to the grounding terminal of the switch. Once the screws are placed properly to hold the metal faceplate onto the switch or receptacle, the metal faceplate will be properly grounded.

OUTDOOR LIGHTING

In some homes, outdoor lighting is installed for year-round use. This means the lamp is mounted on a box or conduit-fed fixture. Some precautions should be kept in mind under these conditions. A single rigid metal conduit may not support a box, even with concrete fill in the ground. Fig. 12-31. The box should be supported by two metal conduits. EMT cannot be used to support a box. It should be supported by rigid conduit. Figure 12-32 shows a number of EMT supporting the box. This is not according to the Code, even though several connections are made to the box. Figure 12-33 shows

307

three of the common violations of the NEC. The box shown is supported by a piece of EMT—violation number one. Only one hub on the box is connected—violation number two. The box is more than 18 inches above the ground—violation number three.

Any outlet box that is mounted on conduit must not be over 18 inches above the ground. It should have two pieces of conduit supporting it by connecting to at least two hubs on the box. EMT should not be used to support a box. It does not have the strength to withstand severe blows.

12-31. *A single rigid metal conduit cannot be used to support a box, even if mounted in concrete.*

12-32. *The use of more than two EMT supports per box is not according to the Code.*

12-33. *Supporting a box with EMT is not according to Code. Having the box over 18 inches above ground is not according to Code. Supporting the box with only one piece of conduit is also a violation of the Code.*

QUESTIONS

1. How far from the floor should convenience outlets be located?
2. At what height should switch boxes be located?
3. What is meant by "fishing" a wire?
4. Do you need to place a GFI in an outside outlet if you are rewiring a house?
5. What is a duplex outlet?
6. How much current does a one-horsepower motor draw? 6.5 amps 240
7. What type of plates should be used to cover outlets installed outdoors?
8. What are the installation requirements for installing ungrounded metal faceplates?
9. Why should metal faceplates not be used on bathroom or kitchen switches?
10. Can EMT be used to support a box outside when it extends past the ground or grade level?
11. How high should an outlet box mounted on right conduit be mounted above the ground?

Appliances, Ovens, Ranges, Cooking Tops, and Air Conditioners

Many modern appliances are portable and use a standard 120-volt circuit with at least 15 amperes. However, some devices use 240 volts and need special plugs and wiring.

ELECTRIC RANGES

The largest appliance in the house is usually an electric range for cooking. It is wired with a permanent connection or with a plug and pigtails. Figure 13-1 shows the plug that connects the range terminals provided by the manufacturer to the surface-mounted box in Fig. 13-2. This plug and receptacle must be capable of handling at least 50 amperes and provide connections for the three wires used in such circuits.

Figure 13-3 shows a flush-mounted outlet for 240 volts. It can be wired in a junction box with the wall plate to fit. Fig. 13-4. In Fig. 13-5 the dryer is shown hooked up properly with a pigtail flexible cord. This type of setup can be used for connecting electric ranges. The 50-ampere line will be No. 6 wire with a red, black, and white lead. The green lead is a ground wire that is connected to the metal part of the dryer, range, or heating unit for the cooking surface.

The plug and outlet shown in Fig. 13-5 are allowed only for convenience. They cannot serve as a switch or as a means of disconnecting the appliance from the line. A separate switching method is necessary to disconnect the circuit from the oven, cooking top, or dryer.

Connecting Ranges Permanently

The National Electrical Code specifies the size of wire that can be used for connecting electric ranges. Feeder capacity must be allowed for household cooking appliances according to Table 220-19 of the Code. This table applies if the appliance is rated over $1\frac{3}{4}$ kilowatts.

13-1. *A pigtail and plug.*

13-2. *A surface-mounted receptacle for 240 volts, 50 amperes.*

13-3. This outlet must be turned to wire. (Slater)

13-4. A faceplate for a flush-mounted single receptable. (Slater)

13-5. Installation of a pigtail. (Sears)

Section 210-19(b), Exception No. 1, refers to Table 220-19 for the sizing of a branch circuit to supply an electric range, a wall-mounted oven, or a counter-mounted cooking unit. Table 220-19 is also used for sizing the feeder wires that supply more than one electric range or cooking unit.

Figure 13-6 shows that the minimum size for a 12 000-watt range would be 35 amperes and the wire would have to be No. 8. The type of wire would be TW—40 A, THW—45 A, XHHW, or THHN—50 A. Number 8 wire is the minimum size to be used for any range rated 8¾ kilowatts or more. The overload protection could be either a 40-ampere fuse or a 40-ampere circuit breaker.

Although the two hot legs of the wire must be No. 8, the exception to the Code says that the neutral conductor can be smaller. See Fig. 13-6. This is primarily because the heating elements of the range will be connected directly between the two hot wires and the maximum current will flow through these wires. The devices that operate on 115 volts will be the oven lamp and the lighting circuits over the top of the range. In some instances, the heating elements for the top portion will be wired to 115 volts at some heat levels.

Minimum hot-leg
rating = 35 A

Range
circuit

Hot

N

Hot

Minimum neutral
rating = 28 A

115 V

230 V

Rated at
12 000 W

13-6. *A 12-kilowatt range with its minimum ampere rating.*

The exception says that the smaller conductor can be not less than 70% of the capacity of the hot wires. In this case, the wire cannot be less than No. 10. The neutral size is figured as follows: 70% of 40 amperes is 28 amperes. Thus, 28 amperes calls for No. 10 wire. The neutral can be No. 10, but the other two wires must be No. 8.

If the range is over $8\frac{3}{4}$ kilowatts, then the minimum size of the neutral conductor must not be smaller than No. 10. Hot legs of the range must not be less than No. 8 wire and the neutral less than No. 10 wire.

The maximum demand for a range of 12-kilowatt rating according to the Table in the Code is 8 kilowatts. This 8000-watt load can be converted to amperes by dividing by 230 volts. This, then, gives 35 amperes. Look up the wire for handling 35 amperes and you find No. 8. Number 8 wire can handle up to 40 amperes.

On modern ranges, the heating elements of the surface units are controlled by five-heat switches. The surface unit heating elements will not draw current from the neutral conductor unless the unit switch is in one of the low-heat settings.

Sizing a Range over 12 kW

If you have a kitchen range rated at 16.6 kilowatts, then the 230-volt supply must be capable of delivering at least 43 amperes.

This is determined by the following:

1. Column A of the NEC Table for *Demand Loads for Household Ranges, Wall-Mounted Ovens, Counter-Mounted Cooking Units and Other Household Appliances Over $1\frac{3}{4}$ kW Rating* states in Note 1 that for ranges over 12 kilowatts and up to 27 kilowatts, the maximum demand in Column A must be increased by 5% for each additional kW (or major faction thereof) above 12 kilowatts.

2. This range is rated at 16.6 kilowatts, which is 4.6 kilowatts more than 12 kilowatts. Column A says to use 5% of the 8000 watts, or 400 watts.

3. The maximum demand for this 16.6-kilowatt range must be increased above 8 kilowatts by using the following: 400 watts × 5 (4 kilowatts + 1 for the .6 kilowatts more than 16.) Therefore, 400 × 5 = 2000. This is added to the 8000 watts used as a base demand and produces 10 000 as the maximum demand load.

4-kW, 115/230-volt
oven (17.4 amperes)

PREWIRED

4 in.
octagonal box

54 in. - $\frac{3}{8}$ in. flexible conduit
w/No. 14 type A wire (Junction
box and leads supplied by
manufacturer)

12/3 NM cable
w/No. 14 bare ground wire
No. 12 ground wire
is attached to
4 in. octagonal box

To 20-ampere
fuse or breaker
in panel

To 40-ampere
fuse or breaker
in panel

8/3 NM
cable

Junction box
on unit
(Neutral
grounded
to unit)

Maximum feeder demand
for both units-8 kW per
column A of Table 220-19
for a single 12-kW range

Note 1: Individual branch circuits supplying
single units are computed at
100 percent demand factor.

Note 2: Equipment grounding
conductors are computed
according to Table 250-95
or Sec. 250-60, whichever
applies. Also Sec. 250-57(a)
permits metallic conduit or
cable armor to be used to
ground fixed equipment.

8-kW, 115/230 volt,
four-burner cook top
(34.7 amperes)

NOT PREWIRED

13-7. *Cooking units with separate circuits.*

4. If you want the amount of current drawn at maximum demand, divide the 10 000 watts by 230 volts (P = E × I, or I = P divided by E). The answer is 43.478 amperes. This is rounded to 43 since the fraction is less that 0.5. Look up the wire size for using a 60°C conductor and you will find the required UL wire for a branch circuit must have No. 6 TW conductors to handle the 43 amperes.

Tap Conductors

Section 210-19(b), Exception 1, of the NEC gives permission to reduce the size of the neutral conductor for ranges or a three-wire range branch circuit to 70% of the current-carrying capacity of the ungrounded conductors. Keep in mind, however, that this does not apply to smaller taps connected to a 50-ampere circuit where the smaller taps must all be the same size. It does not apply when individual branch circuits supply each wall- or counter-mounted cooking unit, and all circuit conductors are of the same size and less than No. 10.

Exception No. 2 of the section previously mentioned allows tap conductors rated at not less than 20 amperes to be connected to a 50-ampere branch circuit that supplies ranges, wall-mounted ovens, and counter-mounted cooking units. These taps cannot be any longer than necessary for servicing the cooking top or oven. See Figs. 13-7 and 13-8.

Note: These units are not prewired types.

6-kW cook top

4-kW oven

3 foot length of 10/3 (Neutral used to ground unit)

Neutral grounds junction boxes

6/3 NM cable, 6/2-8/1 SE cable 6/3 armored cable or 1"C w/2 No. 6 and 1 No. 8

To 50 ampere fuse or breaker in panel

50-ampere circuit (or 40-ampere circuit)

4 11/16 inch square by 2 1/8 inch deep box for 6 No. 6 and 3 No. 10

4-inch square by 2 1/8-inch deep box for 3 No. 6 and 3 No. 10

4 1/2 foot length of 10/3 (Neutral used to ground unit)

13-8. *Using one branch circuit to cooking units.*

Demand for these two units is 8 kW- same as a single range of 12 kW

6-kW 120/240- volt cook top

6-kW 120/240-volt oven

One cook top and two wall ovens with aggregate name-plate ratings not over 15.5 kW may be connected to a 40-ampere branch circuit

Junction box

40-A or 50-A fuse or circuit breaker

13-9. *These two units are treat-ed as one range load.*

No. 8 (40 A) or No. 6 (50 A)

6-kW wall ovens

8000 WATTS + (6x400 WATTS) = 10 400 WATTS

All appliances rated 115/230 V and used in kitchen of residence.

6-kW cook top

Branch circuit

Note 4 of Table 220-19 says that the branch-circuit load for a counter-mounted cooking unit and not more than two wall-mounted ovens, all supplied from a single branch circuit and located in the same room, shall be computed by adding the nameplate ratings of the individual appliances and treating this total as a single range.

That means the three appliances shown may be considered to be a single range of 18-kW rating (6 kW + 6 kW + 6 kW).

From Note 1 of Table 220-19, such a range exceeds 12 kW by 6 kW and the 8-kW demand of Column A must be increased by 400 watts (5 percent of 8000 watts) for each of the 6 additional kilowatts above 12 kW.

13-10. *Separate cooking units on a single circuit branch.*

Figure 13-9 shows two units treated as a single range load. Figure 13-10 shows how to determine branch circuit load for separate cooking appliances on a single circuit according to Section 210 of the Code.

Individual branch circuits can be used. There are some advantages to individual circuits. With this arrangement, smaller branch circuits supply each unit with no junction boxes required. Two additional fuses or circuit breaker poles are required in a panelboard, however. Overall labor and material costs are less than those for the 50-ampere circuit shown in Fig. 13-8. There is a disadvantage though. The smaller circuits will not handle larger units that you may wish to install later. See Fig. 13-11.

In some cases, a single 40-ampere circuit may supply the units. Fig. 13-12. The NEC allows 40-ampere circuits in place of 50-ampere circuits where the total nameplate rating of the cook tops and ovens is less than 15.5 kilowatts. Since most ranges and

6-kW cook top
(26A)

To 30-ampere
fuse or circuit
breaker in panel   10/3 NM cable

4-kW
oven
(17.4A)

To 20-ampere
fuse or circuit
breaker in panel   12/3 w/bare No. 12 ground

13-11. *Smaller wire can be used when separate branches are run for each cooking unit.*

4-kW oven
PREWIRED

54-inch length
*$\frac{3}{8}$ inch flex.

w/3 No. 14
Type A wires

7.5 - kW cook top
PREWIRED

*36-inch length
$\frac{1}{2}$ inch flexible cable

w/3 No. 12
Type A wires

Removed and flexible cable
is connected to junction box

*4-inch oct. box,
1½ inches deep
(See note)

To 40-A
fuse or circuit
breaker in panel   Junction
box

4 x 2$\frac{1}{8}$-inch deep
box may be needed

Neutral is bonded to
each junction box
40-A circuit   8/3 NM cable

2-1/8 x 4 inch square box or
1-1/4 x 4-11/16 inch square box
for 6 No. 8 and 3 No. 14

*Furnished with units   Note: Cubic inch capacity of 1½-by 4-inch octagonal
box is 15.5; according to Table 370-6(b), 3 No. 8
and 3 No. 12 would require 15.75 cubic inches.

13-12. *In some instances it is possible for a single 40-ampere circuit to supply both cooking units.*

combinations of cooktops and ovens are less than 15.5 kilowatts, this can be a very popular arrangement.

DRYER

The electric clothes dryer uses 240 volts for the heating element. It also uses 120 volts for the motor and 120 volts for the light bulb. This means three wires are needed for proper operation. Figure 13-13 shows how the green and white wires are treated in the dryer. Since most home dryers draw about 4200 watts, it is necessary to use a separate 30-ampere circuit with a pullout fuse for disconnecting it from the line. A circuit breaker in the main service panel can also be used to disconnect it from the line. The 30-ampere circuit uses No. 8 wire in most instances. However, it should be noted that some high-speed dryers use about 8500 watts and need a 50-ampere circuit. That would call for a No. 6 wire. There is a difference in the configuration of the receptacle for a 50-ampere device and that for a 30-ampere device. Fig. 13-14. The

GREEN
GROUND WIRE

BLACK

RED

WHITE

MOUNTING BRACKET
GROUNDING LOCATION
WHEN REQUIRED

13-13. *Hookup for an electric dryer.*

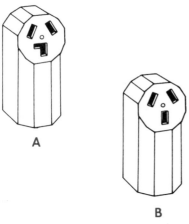

13-14. (A) L-type ground for a 250-volt, 30-ampere receptacle. (B) 250-volt, 50-ampere range receptacle.

13-15. A single-phase, 120-volt plug and receptacle.

surface-mounted receptacle with an L-shaped ground is shown in Fig. 13-14A. It is used with dryers drawing up to 30 amperes. Figure 13-14B shows a surface-mounted receptacle also. It is capable of handling up to 50 amperes at 240 volts. It is used for dryers or ranges. Note the difference in the three outlet connection points. This one has three straight slots.

MICROWAVE OVENS

Microwave ovens are designed to operate on 120 volts. This means they can be plugged into the closest convenient wall outlet. This may become an expensive mistake for many. The microwave oven pulls around 14 amperes. This means there is only 1 ampere of spare capacity left for the circuit. If there is another appliance on the circuit, a circuit breaker or fuse may go. Microwave ovens should have their own circuit. There should be nothing else on the line. They should have a direct connection to the distribution panel. Make sure they are properly grounded as shown in Figs. 13-15 and 13-16.

ELECTRIC HEATERS

There are a number of different designs available to the homeowner today in the field of electric heaters. Heaters such as that shown in Fig. 13-17 may be effective and serve their purpose, but they draw more current than most owners care to admit. They should have a separate circuit to furnish the 120 volts and almost 15 amperes. They have a small electric motor for moving the air, a light bulb for the simulated logs, and a small motor for pulling the device that makes the logs resemble a smoldering fire. The combination of circuitry is shown in Fig. 13-18. Other small heaters should be checked to see how much current they draw. Take the wattage rating and divide by 120 to obtain the current draw of the device.

GREEN WIRE

LUG

WALL
RECEPTACLE

ADAPTER
ASSEMBLY

ATTACH TO
GROUNDED COLD
WATER PIPE
(Remove Paint, etc.)

GROUND WIRE
(Minimum 18 Gauge)

POWER SUPPLY CORD
(with Three-Prong Grounding Plug)

13-16. *Adapting a three-prong 120-volt plug to a two-hole receptacle.*

13-17. *A wall-mounted, decorative, electric heater.*

[120 volts, 60 Hz]

L1   L2

Fan &
motor

Motor

Lite

Heat
limiter

Switch

Thermostat

13-18. *A circuit for Fig. 13-17.*

13-19. *Installation of roof heaters for melting ice and snow.*

ROOF HEATERS AND PIPE HEATERS

In the colder climates it is necessary to heat the eaves to prevent water from backing up under the roofing when the snow melts. These heating elements are installed as shown in Fig. 13-19. Note that the manufacturer has suggested that the heating element be plugged into an outlet such as that servicing the lamp. This is permissible when the heating element is not as long as that shown in Fig. 13-20. The length of the cable determines its wattage rating and, in effect, the current drawn. A separate circuit may be necessary for such heating elements if the current drawn is more than that of the present circuit.

Take a look at Tables 13-A and 13-B to see at what point a new line may be necessary for operation of this heating element. Also keep in mind that adding a line would call for waterproof outlets and proper connection of the cord to the circuit inside a waterproof box.

As can be seen by comparing Tables 13-A and 13-B, it is better to have the 240-volt line supplying power to the heating element. This would, of course, necessitate a separate line to supply the demand. However, if this is known when the house is built, the

13-20. *Installation of roof heaters around a dormer.*

Table 13-A. 110-120 Volt Heating Elements

Length of cable in feet	Average length of roof edge	Wattage	Current (amperes)
20	10	120	1.0
30	15	180	1.5
40	20	240	2.0
60	30	360	3.0
80	40	480	4.0
100	50	600	5.0
120	60	720	6.0
140	70	840	7.0
160	80	960	8.0

line can be added then, rather than later, after construction is complete. Heating elements for water lines exposed to cold temperatures also use the heat produced to prevent frozen, and cracked pipes.

OVERHEAD GARAGE DOORS

Most overhead garage doors for residential use are furnished with a $\frac{1}{3}$-horsepower [248.6-watt] electric motor. This means it draws about 2.07 amperes when running. It does, however, draw much more when starting. This is what dims the lights when it starts. Light bulbs and other electrical devices are sensitive to some high peak voltages.

Every time the garage door motor turns off it gives an inductive kickback. This increased voltage spike produces a shorter life for light bulbs on the same circuit.

Table 13-B. 220-240 Volt Heating Elements

Length of cable in feet	Average length of roof edge	Wattage	Current (amperes)
40	20	240	1.0
60	30	360	1.5
80	40	480	2.0
120	60	720	3.0
160	80	960	4.0
200	100	1200	5.0

13-21. *Wiring of an overhead door opener.*

PORCELAIN RECEPTACLE

COVER PLATE

GREEN LEAD TO GROUND LUG

WIRE NUTS

MOTOR CIRCUIT BREAKER

PORCELAIN RECEPTACLE

13-22. *Connection of a garage door opener to the power line.*

GREEN LEAD TO GROUND LUG

BLACK LEADS

WHITE LEADS

WIRE NUTS

Most manufacturers of garage door openers stress that they are easy to install and can be plugged into the existing lamp socket in the garage. Fig. 13-21. It is suggested, however, that the plug connection to the box shown in Fig. 13-22 be substituted and Romex from a direct line be placed in the garage for operation of the garage door opener.

ELECTRIC WATER HEATERS

Electric water heaters draw current to heat the water. The larger the heater capacity, the higher the current rating of the heating element. They use 240 volts for heating the water.

A double-element water heater is probably better for larger families because it permits a more constant supply of hot water. Double-element heaters have two thermostats. The single-element type has only one thermostat. The size of the elements, the type of thermostats, and the method of wiring for heaters is usually specified by the local power company.

Figure 13-23 shows a typical installation of a hot water heater. Keep in mind that some electric companies offer a special rate for heating water. "Off-peak load" is the term that refers to this special rate. It simply means the power company furnishes electricity during its low load time. The time is already known by the electric company so they can place a meter that measures the use of power for heating water. It also places a time switch on the line so that the usage of electricity is controlled or limited to the time when the power company has a very light demand for power. If you use power at any time other than off-peak time you have to pay at a standard rate, instead of the reduced rate.

As far as the Code is concerned, any fixed storage water heater with a capacity of 120 gallons or less must be treated as a continuous-duty load. This will apply to about 90% of the residential water heaters that use electricity for heating water. The "continuous-duty load" means the ampere rating of the water heater must not exceed 80% of the ampere rating of the branch-circuit conductors.

The only case where the water heater current may load the circuit protective device (circuit breaker or fuse) to 100% of its rating is where the circuit protective device is listed for continuous operation at 100% of its rating. Presently, no standard protective device is rated in this way. Therefore, it is necessary to use the 80% rating as a guide to wire sizing.

13-23. One method of installation of an electric hot water heater. (Sears)

ELECTRIC MOTORS

More electric motors are being used in homes every day. The increased demand for motors for workshop equipment and for water pumps, furnaces, air conditioners, and other devices creates a need for a closer look at circuits that furnish the power for electric motors. Table 13-C shows the wire sizes for individual single-phase motors. It should help with the selection of the proper wire size for workshop use or any other motor application around a residence. Fig. 13-24.

Garbage Disposers

The *Code* emphasizes that the receptacle under the sink must be accessible and located to avoid physical damage to the flexible cord used to connect the motor to the power source. Fig. 13-25.

The cord used for this type of installation should be Type S, SO, ST, STO, SJO, SJT, SJTO, or SPT-3, which is a three-conductor terminating with a grounding-type plug. The cord must be at least 18 and not more than 36 inches long.

The hookup of trash compactors and dishwashers is the same as the hookup for garbage disposers. However, the cord must be from 3 to 4 feet long, instead of $1\frac{1}{2}$ to 3 feet, as for the disposer.

The ampere rating of an individual branch circuit to furnish power to these appliances must not be less than the marked ampere rating of the appliance.

Each of the receptacles serving an appliance with a flexible cord must have some way of disconnecting the circuit so that it may be serviced or repaired by removing the plug from the receptacle. Most appliances need service from time to time. Thus, it is a

Table 13-C. Wire Sizes for Single-Phase Motor Circuits

Horsepower of motor	Volts	Approximate starting current amperes	Approximate full load current amperes	Length of run in feet from main switch to motor								
				Feet	25	50	75	100	150	200	300	400
$\frac{1}{4}$	120	20	5	Wire Size	14	14	14	12	10	10	8	6
$\frac{1}{3}$	120	20	5.5	Wire Size	14	14	14	12	10	8	6	6
$\frac{1}{2}$	120	22	7	Wire Size	14	14	12	12	10	8	6	6
$\frac{3}{4}$	120	28	9.5	Wire Size	14	14	12	10	8	6	4	4
$\frac{1}{4}$	240	10	2.5	Wire Size	14	14	14	14	14	14	12	12
$\frac{1}{3}$	240	10	3	Wire Size	14	14	14	14	14	14	12	10
$\frac{1}{2}$	240	11	3.5	Wire Size	14	14	14	14	14	12	12	10
$\frac{3}{4}$	240	14	4.7	Wire Size	14	14	14	14	14	12	10	10
1	240	16	5.5	Wire Size	14	14	14	14	14	12	10	10
$1\frac{1}{2}$	240	22	7.6	Wire Size	14	14	14	14	12	10	8	8
2	240	30	10	Wire Size	14	14	14	12	10	10	8	6
3	240	42	14	Wire Size	14	12	12	12	10	8	6	6
5	240	69	23	Wire Size	10	10	10	8	8	6	4	4
$7\frac{1}{2}$	240	100	34	Wire Size	8	8	8	8	6	4	2	2
10	240	130	43	Wire Size	6	6	6	6	4	4	2	1

13-24. Electric motors are being used in many appliances throughout the home. (Sears)

good idea to make sure the repair person is properly protected by the removal of the plug from the receptacle. It should have an easily accessible receptacle so the plug can be removed before work begins on the appliance.

The Code permits electric ranges to be supplied by cord and plug connections to a range receptacle located at the rear base of the range. The rule allows such a plug and receptacle to serve as the means for disconnecting the range if the connection is accessible from the front by the removal of a drawer.

Counter top

Sink

13-25. Installation of garbage disposers under the sink. Remember, double-insulated disposers do not require a ground.

Garbage disposer

In some cases, a gas range is used and the clock and oven lights are electric. Then, the receptacle for the lights and clock must be easily accessible without having to disassemble the range.

AIR CONDITIONERS

Electrical wiring for a single-unit air conditioner, window-mounted, is treated as an individual single-motor load of ampere rating as marked on the nameplate.

If the Code is checked, it can be found that, according to Section 430-53(a) two 115-volt air conditioners of 6 amperes each can be used on a 15-ampere circuit. Three 6-ampere conditioners can be used on a 20-ampere circuit if the circuit does not supply any other load. Two 220-volt, 6-ampere units could be operated on a 15-ampere circuit.

Most air conditioners today exceed the 6-ampere full-load current rating. Therefore, the application of more than one air conditioner on a 15-ampere line is limited.

Many local codes avoid the complications of connecting conditioners to existing circuits and connecting more than one conditioner to the same circuit by requiring a separate branch circuit for each conditioner.

If a unit room air conditioner has a fixed (not a cord and plug) connection to its supply, it must be treated as a group of several individual motors. It must be protected in accordance with the sections covering several motors on one branch circuit.

A disconnect is required for every unit air conditioner. The plug and receptacle or a separate connector may serve as the disconnect.

If a fixed unit (no plug and receptacle) is wired into a circuit, the switch or circuit breaker must be readily available to the user. The off switch on the individual unit serves as the disconnect, since it is within reach.

Central Air Conditioners

Some homes use a central air conditioner for cooling. This requires a unit with the compressor and condenser outside. The evaporator is usually located inside. A duct system carries the cool air throughout the house.

Electrical wiring in and to the units varies with the manufacturer. The extent to which the electrician must be concerned with the fuse and circuit breaker calculations depends upon the manner in which the unit motors are fed. It also depends upon the type of distribution system to which the unit is to be connected. A packaged unit is treated as a group of motors. This is different from the approach used for the window units.

Some equipment will be delivered with the branch circuit selection current marked. This simplifies the situation. All the required controls and the size of the wire can be judged according to this information.

In sizing the wire for this type of fixed unit, the selection of components for control must be the "rated-load current" marked on the equipment or the compressor.

The disconnect for a hermetic motor (used in the compressor of the air conditioner) must be a motor-circuit switch rated in horsepower or a circuit breaker.

If a circuit breaker is used, it must have an ampere rating of not less than 115% of the nameplate "rated-load current" or the branch circuit selection current. The larger of the two ratings would be taken as a working base for sizing the wire.

QUESTIONS

1. What is the largest appliance used in the house in terms of electrical load demand?
2. What size wire is used for a 12 000-watt range?
3. What is a tap connector?
4. What is the green wire used for in a dryer pigtail?
5. What is the usual load requirement of a microwave oven?
6. Why is it better to have a 240-volt supply, rather than a 120-volt supply, for a heater?
7. How are most garage door openers connected to the power source?
8. What is used to connect garbage disposals to a power source?
9. What voltage requirements do storage water heaters have?
10. What is meant by continuous-duty load?
11. Why should an appliance served by a flexible cord have some kind of disconnect?
12. How many air conditioners with a 6-ampere rating can you place on a 20-ampere branch circuit?
13. What determines the size of the wire used for a central air conditioning unit?

Knob-and-Tube Wiring

The knob-and-tube wiring method is still in use. It is permitted by Article 324 of the National Electrical Code. It is defined as concealed knob-and-tube wiring. It is a method using knobs, tubes, and flexible nonmetallic tubing for the protection and support of single insulated conductors concealed in hollow spaces of walls and ceilings of buildings.

This type of wiring is permitted to be used only for extensions of existing installations and elsewhere only by special permission under two conditions:

1. In the hollow spaces of walls and ceilings.
2. In unfinished attic and roof spaces that are accessible by stairway or permanent ladder.

There are, of course, some exceptions. Refer to the Code for these exceptions. Splices must be soldered. In-line or strain splices cannot be used. Boxes and switches must comply with various articles of the National Electrical Code.

In the 1950s, the knob-and-tube method was little used, but it still, at that time, had an advantage economically. It was cheaper than using Romex or armored cable. However, today its labor cost alone makes it unthinkable as a method of wiring. It is discussed here so that you will become familar with the method. You may have to add to or remove an existing installation of knob-and-tube at some time.

MATERIALS USED IN KNOB-AND-TUBE WIRING

Conductors

The wires you find in the knob-and-tube method of wiring will probably be the R, RP, RH, or RHT type that is rubber-coated, slow-burning, or weatherproof (SWB). They may also be varnish-cambric (V) or thermoplastic (T or TW) types. In damp locations, conductors should be of the rubber-covered type.

Insulators

Insulators had to be free of projections, sharp edges, or anything that may injure the insulation. Most of the insulators are made of porcelain. Loom, a flexible nonmetallic tubing, was used to protect the electrical conductors. Figure 14-1 shows the loom. Figures 14-2 and 14-3 show practical applications of the loom for protecting the conductors or wires.

Boxes and Other Devices

Boxes and other hardware used in knob-and-tube wiring will be described later in this chapter.

ARTICLE 324. CONCEALED KNOB-AND-TUBE WIRING*

324-3. Uses Permitted. This wiring method is restricted to use only for extensions of existing installations. It is not Code acceptable as a general-purpose wiring method for new electrical work. Under the conditions specified in (1) and (2), concealed knob-and-tube wiring may be used only if special permission is granted by the local inspection authority. See the second sentence of Article 90-4 of the Code.

324-5. Conductors. Conductors for concealed knob-and-tube work may be any of the general-use types listed in Table 310-13 for "dry" locations and "dry and wet" locations such as TW, THW, XHHW and RHH.

324-11. Unfinished Attics and Roof Spaces. Where wires are run on knobs or through tubes in a closed-in inaccessible attic or roof space, the wiring is concealed knob-and-tube work; but if the attic or roof space is accessible, the wiring must conform to Art. 320 as open wiring on insulators.

324-13. Outlet boxes shall comply with Article 370.

324-14. Switches shall comply with Sections 380-4 and 380-10(b).

*Reprinted by permission from NFPA 70-1981, *National Electrical Code,* © 1980 National Fire Protection Association, Boston, Massachusetts

14-1. *A piece of loom used for insulating wiring on knob-and-tube installations.*

14-3. *Using loom to insulate the wires where they enter a receptacle box.*

14-2. *Using loom to insulate at a cross-over point between two wires.*

14-4. *Spacing of wires in an exposed type of knob-and-tube wiring.*

2½ in

3½ in

SPACING THE WIRES

Some people call for 2 inches of spacing for wires running parallel and others call for 3 inches. Figure 14-4 shows wires with 2½-inch spacing. You should keep in mind that all the wiring of earlier days was not subjected to strict inspection or codes. Therefore, the spacing may vary among installations. Wires should be spaced ½ inch from the building surface in a dry location. One inch is needed for spacing in a wet or damp location. In fact, the wet or damp location should not be wired with this method. In a concealed installation, the wires must be separated a distance of at least 3 inches and must be supported at least 1 inch from the mounting surface. The minimum spacing of wires in straight runs and at right-angle turns is shown in Figs. 14-4 and 14-5.

SPACING THE SUPPORTS

When wiring was run over exposed flat surfaces, the knobs and cleats were spaced no further apart than 4 feet, as shown in Fig. 14-6.

A support should have been installed within 6 inches of a wire tap or takeoff. The wire of the tap circuit should always have been secured to this support to insure a strain-free tap.

Rubber-insulated conductors

Split knob

4 ft 6 in maximum

4 in

1 in minimum

14-5. *Concealed installation spacing of wires.*

4 in

Supporting surface

Conductor

1 in

Split knobs for support

14-6. *The spacing of cleats and knobs in knob-and-tube wiring.*

Single-wire cleat for supporting large conductors

Joist

4 ft 6 in maximum

Supports should have been installed within 12 inches of an outlet box. The wires to the box should have been installed loosely so that there was no strain on the terminal connection.

INSTALLATION

Figure 14-7 shows a typical exposed knob-and-tube installation for a hospital unit. It shows the circuiting and the methods of installing the conductors to each of the outlets.

14-7. *Typical installations of knob-and-tube.*

Knobs and Cleats

Split knobs were used to support wire sizes 10 through 14 and could support either one or two wires. They were used as two-wire supports at splices and taps. Figure 14-8A shows the use of split knobs.

Solid knobs were used to support wire size No. 8 or larger. The wires had to be supported on the solid knobs by tying. The conductors used for tying must have the same insulation as the supported conductors. A porcelain solid knob is shown in Fig. 14-8B.

Two or three-wire cleats were occasionally used in supporting wire sizes No. 10 to No. 14. Single cleats had to be used for wire size No. 8 or larger. Cleats were available which supported the wires at distances of $\frac{1}{2}$ to 1 inch from the surface on which the cleats were mounted. Fig. 14-8C.

The installation steps used in mounting the split knobs or cleats are shown in Fig. 14-8C. In the first operation, leather washers, to cushion the porcelain, are threaded on the nails of a two-wire cleat. In the second step, wires are placed in the grooves of the cleat base section and the cleat head and nails are positioned above the wires. The third step shows the cleat in supporting position after the wires were pulled tight and the nails driven firmly into the wood.

One-wire support — Two-wire support

14-8. *Porcelain knobs and cleats. (A) A porcelain split knob supporting one or two wires. (B) A porcelain solid knob with support. (C) Porcelain cleats supporting two parallel wires. Check the sequence from left to right to see the entire procedure for mounting.*

14-9. *Use of the protective tube. (A) The correct and incorrect methods of installing protective tube for a wire cross-over. (B) Protection for a conductor passing over a pipe.*

Tubes

When conductors passed through studs, joists, walls, floors, or partitions, they had to be protected by procelain tubes installed in the hole through the supporting member. These tubes were available in sizes ranging from 1 to 24 inches long and $\frac{5}{16}$-to $1\frac{1}{2}$-inch inner diameter. The tubes had to be long enough to extend through the entire wall. If the wall was too thick to use porcelain bushings, standard iron pipe or conduit had to be used. Bushings were required at both ends of the pipe or conduit. The holes in which the tubes were to be installed were drilled at an angle so that the tube head could be placed on the high side of the hole to prevent it from being dislodged by gravity. The tubes were used to protect wires at crossover points. Fig. 14-9. As the tube was installed on the wire closest to the supporting surface, it was installed on the inner wire, thus preventing the outer wire from making contact with the mounting surface. See Fig. 14-9.

Conductors that passed through timber cross braces in plastered partitions had to be protected by an additional tube extending at least 3 inches above the timber. The extra tubes protected the conductors from plaster accumulation, which collected on the horizontal cross members when plastering was done after installation.

Loom

Loom, a flexible, nonmetallic tubing, was used to support wires where the conductors entered the switch box or made a crossover that could not be supported by knobs or cleats. Loom was made of woven varnished cambric. Support for loom should not be more than 18 inches in spacing. No such loom installation should have run for more than 15 feet. Loom was used for crossovers for insulation purposes. If the wires came

Running boards   Railings

14-10. *Using running boards and railings to protect wiring that may be exposed to damage.*

within $\frac{1}{2}$ inch of support timber, the loom was installed for added protection. It was also used if two wires were closer than $2\frac{1}{2}$ inches. Loom was used on the entry of all boxes. Outlet boxes were designed to clamp the loom securely to the box. See Figs. 14-2 and 14-3.

PROTECTION AGAINST DAMAGE

Running boards were used to protect the conductors from damage in exposed locations. Figure 14-10 shows some examples of running boards and railings used to protect parallel wiring. Exposed wiring located within 7 feet of the floor was considered to be vulnerable to damage. Running boards had to be at least $\frac{1}{2}$ inch thick.

THREE-WIRE INSTALLATIONS

Figure 14-11 shows three-wire possibilities. The installation of wires in groups of three on joists and running boards called for a surface of at least 7 inches wide to make sure wire spacing was $2\frac{1}{2}$ inches. There had to be a spacing of 1 inch for wood clearance beyond each outside wire. When joists were not large enough, one wire was run on top of the joist and the other two wires on the sides.

Concealed three-wire installation consisted of conductors supported in the hollow spaces between walls and ceilings. The wiring was installed in buildings under con-

Top of Running Board

14-11. *Three methods of mounting knobs for support of three-wire circuits.*

Side of Ceiling Joist

Joist Top and Sides

struction after the floors and studding were in place, but before the lathing or any other construction was completed. The wires were attached to devices in boxes that had to have their front edges mounted flush with the finished surface. To facilitate this type of installation, the boxes were generally mounted on brackets or wooden cleats, as shown in Fig. 14-12.

INSTALLATION OF CONSUMING DEVICES

Figure 14-13 shows the procedure used in connecting lighting sockets to open wiring circuits. The base of the porcelain lamp socket was first fastened by wood screws

14-12. *Placing a box in a plastered ceiling.*

to the mounting member. The wires were then stripped of insulation and looped around the screw terminals. Finally, the porcelain head was attached by screwing it onto the base.

Duplex Receptacle

A typical duplex receptacle installation is shown for open wiring in Fig. 14-14. The required knob mounting 12 inches from the box is shown. The placement of loom over the wire at the entry to the box is also shown. The standard mounting height of a receptacle was either 1 foot or 4 feet above the floor, depending upon the location of the outlet.

A   Base

C   Cover attached

B   Wire connection

14-13. *Steps A, B, and C show how a porcelain fitting or socket is mounted to use the parallel wiring of knob-and-tube.*

14-14. *Using the knob-and-tube method wiring to install a duplex receptacle and an exposed type of lamp.*

Lamp Holders

The installation and connection of lampholders used in exposed open wiring are shown in Fig. 14-15. The pigtail socket had permanently attached leads of No. 14 wire or larger. These were paired, but not twisted together unless they were longer than 3 feet. The pendant lampholder was a device to which the lamp cord was attached and from which it was supported by means of an underwriter's knot. Both the pendant and pigtail lampholder sockets were keyless or switchless. They were operated by wall switches to prevent additional strain on the lead wires supporting the sockets.

Knob-and-Tube Service Entrance

Figure 14-16 shows a typical service entrance using the knob-and-tube wiring method. If a service-entrance switch were used instead of a main circuit breaker, a separate fuse cabinet or box would be required to provide individual circuit protection.

A — Elevation — Section A-A — Bottom view

14-15. *(A) Typical wiring to surface receptacle.*

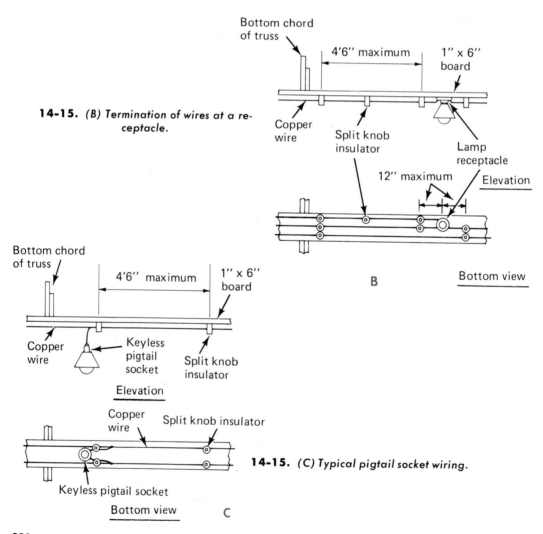

14-15. *(B) Termination of wires at a receptacle.*

14-15. *(C) Typical pigtail socket wiring.*

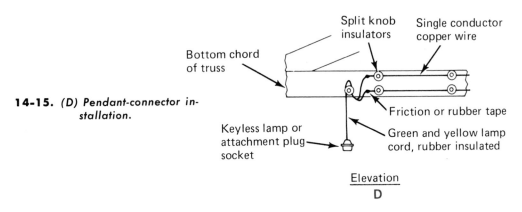

14-15. *(D) Pendant-connector installation.*

Split knob insulators

Single conductor copper wire

Bottom chord of truss

Friction or rubber tape

Keyless lamp or attachment plug socket

Green and yellow lamp cord, rubber insulated

Elevation
D

Wire from the powerline was secured to the building at least 10 feet from the ground for normal installations. When the service entrance was located above a street this height was increased to 18 feet. If the building was not high enough to meet these requirements, the entrance height might be less, provided all conductors within 8 feet of the ground were rubber-insulated. The line wires at the service entrance to a building had to be spaced at least 6 inches apart and should have been supported at least 2 inches from the building by service-entrance insulators or brackets. Upon en-

Flexible fireproof fabric loom

Porcelain unglazed tubes

Inside view of typical end of a building 20 ft wide

Single conductor weatherproof double braid

Glazed porcelain split-knob type insulator

2 x 4 in wiring and circuit breaker support

Panel board circuit breaker

Provide wiring support when more than 4 ft 6 in

14-16. *Circuit breaker installation using the knob-and-tube method of wiring.*

14-17. *Temporary wiring using the knob-and-tube method.*

tering the building, the line wires should have been threaded upward through slanting noncombustible tubes so that moisture would not follow the conductor into the service-entrance switch.

TEMPORARY WIRING

Temporary wiring still exists in older houses and buildings. It was installed to get the job done quickly. It was not intended to be permanent. Some of the methods used in temporary wiring are shown here to familiarize you with them.

A complete installation would require knobs, tubes, cleats, and damage protection. Such an installation would require more time. Generally, the wires were attached to building members with nails, and pigtail sockets were used for outlets.

14-18. *Installation of temporary extension cords.*

Supports

In most cases, the wire sizes were selected in accordance with normal installation. The wires were laid over ceiling joists. They were fastened by nails driven into the joists and then bent over the wire, as shown in Fig. 14-17.

If loom was available, it was used to protect the wire at the nail support. This was very necessary if the wood was wet at the time of installation. It was a good idea to install such temporary wiring at least 7 feet above the floor. Spacing of the wires should have been about the same as for regular knob-and-tube wiring.

Fixture drops are shown in Fig. 14-18. Each tap off the parallel wires should have been individually taped.

This brief explanation of temporary wiring was meant to show you how one job was done quickly. Variations of temporary wiring still exist in millions of older buildings. Many of these temporary wiring systems are still energized and overloaded. Complete rewiring with modern wire and equipment is the only way to make these buildings safe.

QUESTIONS

1. What does knob and tube mean?
2. Is the knob-and-tube wiring method still permitted by the National Electrical Code?
3. Under what conditions can you install knob-and-tube wiring today?
4. At what intervals should the wires be spaced in knob-and-tube wiring?
5. At what intervals are the wires to be supported in knob-and-tube wiring?
6. How do you support wire on solid knobs?
7. What precautions must be taken when conductors are passed through timber cross braces in plastered partitions?
8. What is loom? Where is it used?
9. What is a running board? How is it used?
10. At what intervals should the wires be spaced when three wires are used?
11. How were wires usually attached to building members when wired in temporary fashion?
12. What was used to protect wire when nails were used for temporary wiring security?
13. At what height should temporary wiring be installed?

CHAPTER 15

Making Electricity a Career

The field of electricity offers a variety of career opportunities. Electricians are employed by many different industries—from steel mills to drug manufacturers, from coal mines to the merchant marine. The residential electrician is a special type of electrician. He or she works with the wiring of homes. Today, many modern appliances and electrical devices enable us to lead more comfortable lives. To operate properly, these devices need to have the correct voltage and current. This is the job of the electrician—to wire homes safely and in accordance with Code standards. This ensures that electrical equipment will operate properly and efficiently.

INDUSTRIES THAT EMPLOY ELECTRICIANS

Many industries which rely heavily on electrical equipment hire full-time electricians to keep that equipment in good working order. For example, the aluminum industry needs electricians to install and repair electrical fixtures, apparatus, and control equipment. In the iron and steel industry, electricians install wiring and fixtures and hook up electrically operated equipment. Electrical repairers (motor inspectors) keep the wiring, motors, switches, and other electrical equipment in good working condition. The paper industry employs electricians to repair wiring, motors, control panels, and switches. Fig. 15-1.

15-1. Industries of every type require electricians to service electrical equipment, including control panels. (Courtesy U.S. Department of Labor)

15-2. *The operation of delicate monitoring devices in a hospital depends on the proper maintenance of the hospital's electrical systems. (Courtesy U.S. Department of Labor)*

Railroads employ electrical workers to install and maintain the wiring and electrical equipment in locomotives, cars, and railroad buildings. Some workers also lay and maintain power lines.

Ships at sea need their own crew of maintenance workers on board. In the merchant marine industry, the ship's electrician repairs and maintains electrical equipment, such as generators and motors. The electrician also tests wiring for short circuits and removes and replaces fuses and defective lights.

In hospitals, the lives of patients often depend on equipment run by electricity. Proper maintenance and fast repair of electrical equipment are vital. Many hospitals, therefore, employ full-time electricians. Fig. 15-2.

Factories that make medicines also rely on electricians to keep equipment running properly and prevent costly breakdowns. Electricians working for these companies also install and repair the various types of electrical equipment used.

Electricians are involved in the coal mining industry. They check and install electrical wiring in and around the mines. In addition, they help repair and maintain the machinery used for mining.

Today's cars have a number of electrically operated devices and systems. In the auto industry, electrical engineers design the car's electrical systems, such as the ignition system, lights, and accessories.

In the construction industry, electricians assemble, install, and wire systems for heat, light, power, air conditioning, and refrigeration. Electrical inspectors then check

15-3. *The electrical engineer is a skilled professional who sometimes helps design facilities that distribute electric power. (Courtesy U.S. Department of Labor)*

15-4. *For a career in teaching, the electrical engineer usually needs a graduate degree. (Courtesy U.S. Department of Labor)*

these systems to make sure that they work properly and comply with electrical codes and standards. The inspectors visit worksites to inspect new and existing wiring, lighting, sound and security systems, and generating equipment. They also check the installation of the electrical wiring for heating and air conditioning systems, kitchen appliances, and other components.

CAREERS IN ELECTRICITY

Although each industry has its own special requirements, careers in electricity can be divided into three general categories: electrical engineer, construction electrician, and maintenance electrician.

Electrical Engineer

Electrical engineers design, develop, and supervise the manufacture of electrical and electronic equipment. The projects they work on include electric motors and generators; communications equipment; electronic equipment such as heart pacemakers, pollution-measuring devices, radar, computers, lasers, and missile guidance systems; and electrical appliances of all kinds. They also design and operate facilities for the generation and distribution of electric power. Fig. 15-3.

Usually, electrical engineers specialize in one major area, such as electronics, computers, electrical equipment, or power. Within these areas, there are still more specialized fields, such as missile guidance and tracking systems.

About 300 000 people are employed as electrical engineers. Most work for makers of electrical and electronic equipment, aircraft and parts, business machines, and professional and scientific equipment. Many work in the communications industry and for electric light and power companies. Electrical engineers also work for government agencies, for construction firms, for engineering consultants, or as independent consulting engineers. Others teach in colleges and universities.

A bachelor's degree in engineering is required for most beginning engineering jobs. In a typical four-year curriculum, the first two years are spent studying basic sciences (mathematics, physics, introductory engineering) and the humanities, social sciences, and English. The last two years are devoted mostly to specialized engineering courses.

Graduate training is now required for many jobs, particularly for teaching and for jobs in specialized areas. A number of colleges and universities now offer five-year master's degree programs. Fig. 15-4.

All fifty states and the District of Columbia require licensing for engineers whose work may affect life, health, or property, or who offer their services to the public. To obtain a license, an engineer usually must have a degree from an accredited engineering school, four years of related work experience, and must pass a state examination.

Engineering graduates usually begin work under the supervision of experienced engineers. With experience and proven ability, they advance to positions of greater responsibility.

Electrical engineers should be able to work as part of a team. They should have creativity, an analytical mind, and a capacity for detail. They should be able to express their ideas well, both orally and in writing.

The employment outlook for electrical engineers is good. Increased demand for computers, for electrical and electronic consumer goods, for military electronics, and for communications and power generating equipment will increase the need for engineers in these fields.

Construction Electrician

As stated earlier, construction electricians assemble, install, and wire systems for heat, light, power, air conditioning, and refrigeration. They also install electrical machinery, electronic equipment, controls, and signal and communications systems.

Construction electricians must be able to read blueprints and specifications. For safety reasons, they must follow the regulations in the National Electrical Code. Their work must also conform to local electrical codes.

There are about 250 000 construction electricians in the United States. Most work for electrical contractors or are self-employed as contractors. Some work for government agencies and businesses that do their own electrical work.

Training for a career as a construction electrician usually involves a four-year apprenticeship program. Apprenticeship programs are sponsored and supervised by local union-management committees. The programs include both classroom and on-

15-5. *The residential, or construction, electrician installs the wiring in residential and commercial buildings. (Courtesy U.S. Department of Labor)*

the-job training. To qualify for an apprenticeship, an applicant usually must be a high school or vocational school graduate. Courses in electricity, electronics, mechanical drawing, science, and shop provide a good background.

Physical strength is not essential, but skill with one's hands, agility, and good health are important. Good color vision is essential because electrical wires are often identified by color.

In most cities a license is required for employment. The electrician must pass a test which requires a thorough knowledge of the craft and of state and local building codes.

Employment outlook is good. The number of construction electricians is expected to grow faster than the average for all occupations. In any given year, the number employed depends on the amount of construction activity. When jobs are not available, however, construction electricians may be able to transfer to related jobs. For example, they may work as maintenance electricians.

Maintenance Electrician

Maintenance electricians keep lighting systems, transformers, generators, and other electrical equipment in good working order. They may also install new equipment.

The duties of maintenance electricians depend on where they are employed. Electricians working in large factories may repair particular items such as motors and welding machines. Those in office buildings and small plants usually fix all kinds of electrical equipment. Regardless of location, electricians spend much of their time doing preventive maintenance. They check equipment to locate and correct defects before breakdowns occur.

Nearly 300 000 people work as maintenance electricians. More than half are employed by manufacturing industries. Many others work for utilities, mines, railroads, or the government.

Most maintenance electricians learn their trade on the job or through apprenticeship programs. The apprenticeship programs and the requirements for getting into the programs are the same as those for construction electricians. Hand skills, agility, and good health are important. Good color vision is required.

Because of increased use of electrical and electronic equipment by industry, the demand for maintenance electricians will increase. Growth in the number of job openings is expected to be steady. The demand for maintenance electricians is not very sensitive to ups and downs in the economy.

Residential Electrician

Residential electrician is another name for the construction electrician. He or she is usually an expert in house wiring. The wiring used for homes is becoming more complex every day. Today, an electrician needs more than on-the-job training. Apprenticeship programs can provide valuable training. Your local union will have an apprenticeship program. Your local vocational school can also help. Fig. 15-5.

INFORMATION SOURCES

• For information about a career as a construction electrician, contact local electrical contractors, a local union of the International Brotherhood of Electrical Workers, a local union-management apprenticeship committee, or the nearest office of the state employment service or state apprenticeship agency. Some state employment service offices screen applicants and give aptitude tests.

• Information about apprenticeships or other work opportunities for maintenance electricians is available from local firms that employ maintenance electricians and from local union-management apprenticeship committees. In addition, the local office of the state employment service may provide information about training opportunities. Some state employment service offices screen applicants and give aptitude tests.

• For general information about the work of electricians, contact:

International Brotherhood of Electrical Workers
  1125 15th Street N.W.
  Washington, D.C. 20005
National Electrical Contractors Association
  7315 Wisconsin Avenue
  Bethesda, MD 20014
National Joint Apprenticeship and Training Committee for the Electrical Industry
  1730 Rhode Island Avenue N.W.
  Washington, D.C. 20036

Index

A

Above-ground distribution system, 66
Adapters, grounding, 95-99
Air conditioners
  circuits for, 186-187
  general discussion, 326
Alarms. *See also* Detectors
  burglar, 189-192, 248-249
  fire, 248-249
  security, 249
Alphabet, Greek (table), 19
Alternator. *See* Generator, AC
Aluminum connections, 220-221
American Insurance Association, 108
Ammeter, 19, 29
Ampere, explained, 18-19
Apartment complex, service installation for, 76
Appliances, electric, 310-327
Apprenticeship programs, 345-347
Area, cross-sectional, and resistance, 20
Armature, 44
Army Corps of Engineers, 68
Atom, 11-12
Attic, accessible, wiring of, 140-141
Awl, 152

B

Ballast, dimming, 254-255
Basements, wiring of, 140-141
Bathrooms, wiring of, 134-136
Batteries, 16
Bedrooms, wiring of, 133-135
Bell, fire alarm, 189-192
Bit, drill, 145-146, 155-158
Blanks, snap-in, 274
Bonneville Power Administration, 68
Box
  handy, 202-205
  installing, 293-294
  secure, installing, 302-303
Box finder, 199
Box volume, and Romex, 196-197
Brace, bit, 145-146, 155
Brushes, 44
Btu content, 48
Bureau of Reclamation, 68
Bushings
  for BX cable, 200
  conduit, 282-283
Butane, 48-49
BX cable, 197

C

Cable. *See also* Conduit
  AC, 262-290
  armored (AC), 111-114, 257-290
  BX, 197-209, 257-290
  Romex, 171, 196-225
  sealtite flexible, 288-290

  snow-melting, 143
  thermostat, 245-248
Cable tracer, 199
Canadian Standards Association (CSA), 109-110
Canopies, 306-307
Carports, wiring of, 143
Career opportunities, 342-347
Cell
  hydrogen-oxygen, 18
  lead-acid, 17
  oxygen-concentration, 17
  photoelectric, 16
Chalk line, 151
Chemical action, and electricity, 16
Chimes, door, installing, 192-193
Chisel, 146-147, 154
Circuit breaker, 36-37, 100, 169-171, 222-223, 337-339
Circuit-breaker box, 223-224
Circuit load, figuring, 123-125
Circuit protective devices, 222-224
Circuits
  air conditioning, 186-187
  branch, 174-184
  defined, 18
  explained, 22-24
  house, 160-195
  low-voltage, 226-248
  relay, 37
  space heating, 184-186
Circuit tester, 147
Clamps
  in boxes, 219
  pipe, 281-282
Clearances, service drop, 265-266
Cleats, porcelain, 332
Climate control, 245
Closets, wiring of, 138
Coal, 56-59
Code, National Electrical (NEC)
  and air conditioners, 326
  and BX cable, 262
  and conduits, 283-288
  and electric ranges, 325
  and garbage disposers, 324
  general discussion, 106-122
  and knob-and-tube wiring, 329
  and water heaters, 323
Coil
  current, 32
  field, 44
  potential, 32
Color code, resistor, 38-39
Communications, wiring for, 188
Commutator, 44
Compounds, 11
Conductor length, and resistance, 20
Conductors. *See also* Wire

general discussion, 13-14
in knob-and-tube wiring, 328
number of, in a box, 216-220
number of, in conduit, 284
protection of (in knob-and-tube wiring), 334
tap, 313-317
Conduit. See also Cable
flexible metal, 121-122, 287-290
installing, 264-290
intermediate metal (IMC), 283-285
nonmetallic, 120
polyvinyl chloride (PVC), 284-286
rigid metal, 114-120
Conduit bender, 146, 149, 274-275
Conduit stop, 281
Connectors. See also Fittings, conduit
for BX cable, 198-199
cable, 258-261
conduit, 278-281
Control units, for remote control, 232-235
Cook tops, 313-317
Cords, temporary, 81
Consuming devices, installing in knob-and-tube wiring, 336-340
Coulomb, 18
Couplings, conduit, 278-281
Cover plates, 205-206
Covers, 306-307
Cow milkers, electric, 69
CSA. See Canadian Standards Association
Current
explained, 18-19
fatal, 82-87
and Ohm's law, 24-25
Current flow, 13
Cutter
BX, 257-258
Roto-Split®, 111-113

D

Detectors. See also Alarms
fire-heat, 189-192
smoke, 191-192, 248-249
Dimmers, electronic, 249-252
Dining area, wiring of, 132-133
Distribution panels, 169-171
Door bells, installing, 192-193
Door chimes, 245-247
Door opener, 245, 247
Drill, electric, 150
Dryers, 317-318

E

Edison, Thomas A., 67, 106
Elbows, for conduit, 268-271
Electricians, 342-347
Electricity
controlling, 32-37
current, 11
distribution of, 61-78
general discussion, 11-41
generating, 15-18
and magnetism, 43-46

measuring, 27-32
producing, 43-60
in rural areas, 68-75
static, 11
Electrodes, 17
Electrodynamometer, 32
Electrolyte, 17
Electromagnets, 36
Electromotive force (EMF), 18
Electrons, 12-13
EMF. See Electromotive force
EMT. See Tubing, electrical metallic
End caps, 216
Energy, 13
Energy sources, mechanical, 46-54
Engine
diesel, 47-52
gas, natural, 48-52
gasoline, 46-52
Engineer, electrical, 344-345
Entrance caps, 270-271
Entrance ell, 271
Entrance signals, 187-188
Environmental Protection Agency (EPA), 56
EPA. See Environmental Protection Agency
Extension cords
inspection of, 88-89
and shock hazards, 87-89
temporary, 340-341
Exterior entrances, wiring of, 130
Exterior grounds, wiring of, 143-144

F

Faceplates, metal grounding of, 306-307
Family room. See Recreation room
Federal Power Commission, 68
First aid, for electrical shock, 85
Fish tape, 148-149
Fish wire, 148-149
Fission, 54
Fittings, conduit, 268-290. See also Connectors
Fixtures, lighting, 209-211
Fixture stems, 273
Flex. See Conduit, flexible metal
Flexsaw, 153-154
Friction, and electricity, 15
Fuse puller, 151
Fuses, 100, 222

G

Gage, standard wire, 149-150
Garage doors, overhead, 321-322
Garages, wiring of, 143
Garbage disposers, 324-326
Gas
bottled, 48-49
natural, 48, 56-59
synthetic, 58
Generators
AC, 44-46. See also Alternator
exotic, 17-18
fossil-fuel power, 56-60
magnetohydrodynamic (MHD), 18

GFI. *See* Ground-fault interrupter
Gouge, 154
Greenfield. *See* Conduit, flexible metal
Ground-fault interrupter, 100-105, 130
Grounding
 general discussion, 92-105, 176
 of metal faceplates, 306-307
 of a receptacle, 202-204
 of Romex, 200
 of a switch box, 295-297
 of systems, 167
Grounding adapter, 212
Grounding clip, 200
Grounding connection, 214
Grounding jumper, 219
Ground wire, attaching to box, 176
Ground wires, in connecting wires in boxes, 218-219

H

Hacksaw, 145-146, 153
Halls, wiring of, 138-139
Hammers, 145-146
Handsaw, 150
Heat, and electricity, 16
Heaters
 electric, 318-319
 pipe, 320-321
 roof, 320-321
 water, 322-323
Heat gain, allowable (table), 187
Heat loss values (table), 185
Horsepower, 25-26
Household equipment, conductor capacities for (table), 172-173
Hubs, rigid conduit, 274-275

I

IMC. *See* Conduit, intermediate metal
Indenter tool, 278-279
Insulators
 general discussion, 14
 in knob-and-tube wiring, 328
Intercoms, 245-246
International Brotherhood of Electrical Workers, 347

J

Jackknife, 146, 148
Joule, defined, 25
Jumper, bonding, 98-99

K

Kilowatt-hour, 26
Kilowatts, 26
Kirchoff's law of voltages, 22-23
Kitchen, wiring of, 136-138
Knobs, porcelain, 332

L

Ladders, 81-82
Lamp holder, in knob-and-tube wiring, 337
Lampcord, replacing, 211-212

Lamps
 fluorescent, controls for, 255-256
 incandescent, controls for, 248-254
 pilot, 236-237
Laundry area, wiring of, 138
Lavatories, wiring of, 134-136
Light, and electricity, 16
Lighting
 outdoor, 307-308
 temporary, 79-81
Lighting controls, 248-256
Living room, wiring of, 131-132
Location
 of outlets, 124
 of switches, 124
Locknuts, 273-274
Loom, 328-329, 333-334
Lugs (in splicing), 213

M

Magnetism, and electricity, 16, 43-46
Male enlargers, 272-273
Material(s)
 electrical, 13-15
 and resistance, 20
Merrill, William Henry, 107
Meter movement, D'Arsonval, 27-28
Meter pole, farm, 71
Meters
 general discussion, 27-33
 kilowatt-hour, 72-75, 193-195
Metric system of measurement
 conversion tables, 40-41, 158
 general discussion, 11
Microammeter, 19
Milliammeter, 19
Milliwatt, 26
Mobile home, electrical installation for, 75-77
Molecule, 12-13
Motors, electric, 324-326
Mud ring, 219
Multimeter, 28-30

N

Nail puller, 158
National Board of Fire Underwriters, 108
National Electrical Code (NEC), wiring according to, 110-122
National Electrical Contractors Association, 347
National Electrical Manufacturer's Association (NEMA), 89
National Fire Protection Association, 106-107
National Joint Apprenticeship and Training Committee for the Electrical Industry, 347
NEC. *See* National Electrical Code
NEMA. *See* National Electrical Manufacturer's Association (NEMA)
Neutrons, 12-13
New York Board of Fire Underwriters, 106
Niagara Power Project, 53-54
Nipples, 272
Nucleus, 12-13

INDEX

O

Ohm, 20
Ohm, Georg S., 24
Ohmmeter, 30-31
Ohm's law, 24-25
Outlets
  adding, 292-298
  air conditioning, 187
  location of, 124
  number necessary, 125-144
  space heating, 187
  weatherproof, 130
Ovens, microwave, 318

P

Patios, wiring of, 143
Permits, 163
Piezoelectrical effect, 16
Pigtail, 310-311
Pilot light assembly, 235
Plans, wiring, 163
Pliers, linesmen's. See Side cutters
Plugs
  electric, general discussion, 89-95
  three-prong, adapting, 319
Poles, field, 44
Porches, wiring of, 142
Potential drop, 18
Power
  general discussion, 25-26
  single-phase, 70
Power companies, 61
Power formulas, 26
Power plants, AC, 45-46
Power plants, nuclear, 54-56
Precipitator, electrostatic, 15
Pressure
  electrical, 18
  and electricity, 16
Preventive maintenance, of extension cords, 88-89
Propane, 48-49
Protons, 12-13
PVC. See Conduit, polyvinyl chloride

R

Radio, wiring for, 188
Ranges, electric, 310-317
REA. See Rural Electrification Administration
Reactor, nuclear, 54-56
Receptacles
  duplex, in knob-and-tube wiring, 336-337
  outside, installing, 303
  and safety, 89-95
Recreation room, wiring of, 139-140
Rectifiers, in remote control circuits, 230-231
"Red head," 262
Regulations, local, 163-165
Relay control center, 234
Relays, 37, 226-227
Remote control, low-voltage, 226-248
Resistance
  general discussion, 14, 19-20

and Ohm's law, 24-25
Resistors, 37-41
Riser support, service entrance, 73
Romex. See Cable, Romex
Romex stripper, 146, 149
Roosevelt, Franklin D., 69
Rule, folding, 145-146
Rural Electrification Act, 69
Rural Electrification Administration (REA), 68-70

S

Safety, general discussion, 79-105, 244
Safety devices, 99-105
Saw
  general discussion, 153-154
  hole, 150-151
  keyhole, 145-146
Scaffolding, 82-83
Screwdrivers, spin-tite, 150-151
Semiconductors, 14-15
Service
  house, 160-195
  150-ampere, 173
Service capacities (table), 162
Service drop, 164-165, 169, 264-265
Service entrance
  general discussion, 161-169
  knob-and-tube, 337-340
Service installation, electrical, 70-77
Shock, electric, 82-87
Shunts, 29
Side cutters, 147
Socket, pigtail, 337-338
Soldering iron, 152
Solenoids, 35-37
Space heating, circuits for, 184-186
Splicing, of wires, 212-216
Stator, 32
Stairways, wiring of, 139
Straps, pipe, 281-282
Studs, locating, 293-294
Supports
  for flexible conduits, 287-288
  knob-and-tube spacing, 330-331
Switch boxes, for Romex, 200-224
Switches
  in branch circuits, 174-184
  extension, 236
  general discussion, 33-35
  in-line, 211
  location of, 124
  in low-voltage remote control, 228-230
  outside, installing, 303
  photoelectric-control, 252-253
  push-button, replacing, 291-292
  wall, adding, 295-296
Symbols, line voltage, 243

T

Tables
  allowable heat gain, 187
  appliance wattages, 173

351

boxes and maximum number of wires, 217
conductor capacities for household equipment, 172-173
conduit and tubing dimensions, 118
copper wire, 21
customary-metric conversions, 40-41
effects of electric current, 84
flexible cord ampacities, 86
flexible cord types, 87
Greek alphabet, 19
human resistance to electric current, 86
maximum heat loss values, 185
maximum number of conductors in conduit, 116-118
metric measurements, 158
110-120 volt heating elements, 321
plug and receptacle voltage and current combinations, 90-92
portable tool extension cords, 88
properties of condutors, 119
rigid conduit and intermediate metal conduit, 285
rigid nonmetallic conduit support, 285
safe distance from live circuits in air, 86
service capacities and house sizes, 162
symbols for residential wiring plans, 129
220-240 volt heating elements, 321
volume required for each wire size, 218
wire sizes for single-phase motor circuits, 324
Telephone lines, 70
Television, wiring for, 188
Temperature, and resistance, 20
Tennessee Valley Authority (TVA), 61, 68
Terraces, wiring of, 143
Tester
  circuit, 292-293
  fuse, 292
  ground continuity, 198
Testing, of circuits, 292
Test light, 145-146
Thermistors, 20
Thermocouple, 16
Three-wire installations, 334-335
Toolholder, electrician's, 152
Tools
  basic, 145-158
  multi-purpose, 145-146
Torch, propane, 152
Transformer, 227, 248-250
Troubleshooting, of remote-control wiring, 244-245
Tube, protective, 333

Tubing, electrical metallic (EMT), 121, 275-284
TVA. See Tennessee Valley Authority

U

UL. See Underwriters' Laboratories
Underground distribution systems, 64-68
Underground service, 74
Underwriters' knot, 211
Underwriters' Laboratories (UL), 82, 107-109
Underwriters' Laboratories (UL) label, 82
Utility rooms, wiring of, 140-141

V

Ventricular fibrillation, 84-85
Voltage
  explained, 18-19
  and Ohm's law, 24-25
Voltage drop, 18
Voltmeter, 19, 29-30

W X Y Z

Wall plates, 235-236
Warning systems, household, 189-192
Watt, defined, 25
Wattages, appliance (table), 173
Wattmeter, 31-32
Wire. See also Conductor
  aluminum, 20, 221-222
  copper, 20-21
  fishing of, 300-301
  in remote control circuits, 231
  running through the floor, 299-301
  spliced, in counting wires, 219
  splicing, 212-216
Wire cutter, 149
Wire nuts, 215
Wire size, 20-21, 172
Wire stripper, 149
Wiring
  of appliances, 310-327
  in an attic, 301-302
  with conduit, 257-290
  for house service, 160-195
  knob-and-tube, 110, 328-341
  of lighting controls, 248-256
  low-voltage, 245-248
  remodeling, general discussion, 291-309
  residential, general requirements, 124-144
  temporary, 79-80, 340-341
  underground, 304-306
Wrench, lever-jaw, 146-147
Wrenches, spin-tite, 150-151